Integrated Circuits

ACS SYMPOSIUM SERIES **290**

Integrated Circuits: Chemical and Physical Processing

Pieter Stroeve, EDITOR

University of California—Davis

Developed from the Winter Symposium sponsored by
the Division of Industrial and Engineering Chemistry
of the American Chemical Society,
University of California—Davis,
March 26–27, 1984

American Chemical Society, Washington, D.C. 1985

Library of Congress Cataloging in Publication Data

Integrated circuits.
 (ACS symposium series, ISSN 0097-6156; 290)

 Includes bibliographies and indexes.

 1. Integrated circuits—Very large scale integration—
Design and construction—Congresses.

 I. Stroeve, Pieter, 1945-   . II. American Chemical
Society. Division of Industrial and Engineering
Chemistry. III. Series.

TK7874.I5462   1985   621.381'73   85-19944
ISBN 0-8412-0940-5

ACS Symposium Series

M. Joan Comstock, *Series Editor*

Advisory Board

FOREWORD

The ACS SYMPOSIUM SERIES was founded in 1974 to provide a medium for publishing symposia quickly in book form. The format of the Series parallels that of the continuing ADVANCES IN CHEMISTRY SERIES except that, in order to save time, the papers are not typeset but are reproduced as they are submitted by the authors in camera-ready form. Papers are reviewed under the supervision of the Editors with the assistance of the Series Advisory Board and are selected to maintain the integrity of the symposia; however, verbatim reproductions of previously published papers are not accepted. Both reviews and reports of research are acceptable, because symposia may embrace both types of presentation.

CONTENTS

vii

PREFACE

THE MANUFACTURE OF INTEGRATED CIRCUITS has become a sophisticated technology involving complex physical and chemical operations, often under computer control, that are designed to build structures or modify properties of materials in very thin films. The complexity of any manufacturing process can be understood by studying the individual unit processes used to manufacture integrated circuits. The fundamental physical and chemical phenomena occurring in these unit processes are under intensive investigation by scientists and engineers. This fact is somewhat surprising because the unit processes are used extensively in processing. However, a further understanding of the basic processes may lead to better process control and consequently to better product yield and quality. The fundamental phenomena occurring in the unit processes include mass transfer, heat transfer, fluid mechanics, chemical reaction, phase transformations, and adsorption. To be able to predict how these processes influence the characteristics of the electrical components built on the wafer's surface, one also must have a good background in solid state physics, materials science, and electrical engineering. Therefore, research in the area of integrated circuits has become interdisciplinary, and researchers come from such distinct fields as chemical engineering and solid state physics.

The chapters in this volume present a cross section of some of the most important areas of study in the fabrication of integrated circuits. Many of the chapters are reviews, and the book gives an overview of many areas important in the manufacture of integrated circuits. Both experts and novices will find material in this text that will increase their knowledge of the phenomena occurring in the unit processes.

PIETER STROEVE
University of California—Davis

April 15, 1985

Unit Processes in the Manufacture of Integrated Circuits

Pieter Stroeve

Chemical Engineering Department, University of California, Davis, CA 95616

The fabrication of integrated circuits is very complex. The intricacy of a particular process can be
understood by considering the individual unit processes used to make the product. Key unit processes
are consistently used in the manufacture of integrated circuits. An improved understanding of the
basic physical and chemical fundamentals of each process will help the manufacturing engineer in controlling the uniformity of the product or to design
new processes.

The microelectronic revolution is the term utilized for the phenomena of manufacturing large numbers of electronic components on
thin silicon chips and the utilization of these chips in sophisticated electronic devices. The push for miniaturization came in
part because of economic needs to produce circuits at much lower
cost and at lower power requirements, the development of small
electronic devices, and the need to produce high speed electronic
computers. The microelectronic revolution began with the inven-
tion of the transistor in the late 1940's and has grown to a
multi-billion dollar industry today. The growth of the industry
is truly amazing considering that the first integrated circuits
were marketed in early 1960's. At that time, integrated circuits
were "small-scale" in the sense that about 10 components would be
present per single chip. In the mid to the end of the 1960's
medium-scale integrated circuits were build containing up to 1024
components per circuit, or a 1K bit circuit. In the 1970's large-
scale circuits were built and in the 1980's we now speak of very
large scale integrated circuits (VLSI). Presently, circuits are
built with 262,144 components per circuit or 262K, which is equal
to 2^{18} components on a single chip. In the near future, 1M bit
chips will be a standard product. This number is staggering

0097–6156/85/0290–0001$06.00/0
© 1985 American Chemical Society

considering that a chip is approximately 0.5 by 0.5 cm. Because of the larger number of components, the cost per component is a small fraction of a penny ($<0.01\cent$), and the basic cost per chip continues to decrease even though the number of components per chip have increased.

The design of 262K and 1M bit chips is a very complicated process. Computer-aided design techniques are now used to perform many of the time-consuming tasks such as the relative placement of transistors, resistors and diodes, and the design of interconnections. Large computer programs and data bases are required to perform such design calculations. The computer-aided design techniques can be used to verify that the design conforms to the constraints imposed by the integrated circuit fabrication process.

The manufacture of integrated circuits involve a series of lithographic, deposition, and etching steps to fabricate patterned layers. This process is often called planar processing or planar technology.

The manufacture of thin film products does not only apply to the making of integrated circuits but also magnetic bubble memories, thin film recording heads, tapes and disks. The physical and chemical processes carried out involve heat and mass transfer, momentum transfer, surface phenomena, high temperature chemistry, radiation, etc. Although the material discussed in this book focuses on the manufacture and fundamentals of integrated circuits, many of the basics are applicable to the fabrication of other thin film devices.

Planar Technology

Wafer processing is often called planar processing or planar technology because small and thin planar structures are built on thin wafers ($\sim$ 500 μm thick) of ultra pure silicon or germanium or any other suitable semi-conductor material. The thin wafers are cut from a rod of pure material, which is a single crystal, and then polished. The structures built on the surface of the wafer are electrical components such as resistors, capacitors, diodes, junction transistors, MOSFET transistors, etc. Each wafer contains 200 to 500 chips, with each chip identical to the others.

In any wafer fabrication process, the surface region of the wafer undergoes a number of processing steps involving changes in chemical composition and physical state, as well as the controlled deposition of material on the wafer's surface. Typically, the thickness of the layers of material may be from 0.1 μm to 1 μm thick while the width and length of the characteristic features within the layers, which define the dimensions of each electrical component, are a few microns. For 1 Megabit RAMs the dimensions are now less than 1.7 μm. The control of the processing steps must be extremely precise in order to make a uniform product from chip to chip. A simple example of the manufacture of a component is the capacitor. A capacitor can be made by locally oxidizing the semiconductor substrate, which is silicon here, to silicon dioxide. Whereas silicon is a semiconductor, silicon dioxide is an excellent insulator. If the thin layer of SiO_2 separates a conductor, such as a metal contact, from the semiconductor, then a capacitor is formed. Electrical contact to the semiconductor material can be made in a variety of ways, e.g., by etching an opening in the silicon dioxide and providing a metal contact there. Transistors are more complex in structure than capacitors and depend on the particular design of the transistor. Figure 1 shows a cross-section of a bipolar npn transistor which uses silicon dioxide as an isolator to separate it from other electrical components on the wafer's surface. There are, of course, a variety of ways to isolate electrical components so that the variety of different designs of components is numerous. As a consequence, the fabrication processes are also very different.

Unit Processes in Fabrication

In any wafer fabrication process, the surface region of the wafer undergoes a large number of lithographic, etching, deposition and other physical and chemical processes. The complexity of a process for a given type of chip is impressive. In the training of process engineers, it makes little sense to study the whole manufacturing process for each type of product, since any manufacturing process can be divided into a series of steps or unit processes. The basic principles of each unit process are for the most part independent of the material or the characteristics of the system in which the process is carried out. In the understanding of an integrated circuit manufacturing process, each unit process used to manufacture the product can be studied individually. In chemical engineering processes, the unit processes are often called "unit operations". Traditionally, in chemical engineering, the restriction to the definition of unit operations is that the changes induced in the material are mainly physical (1). In the manufacture of integrated circuits, many of the unit processes involve mainly chemical changes so perhaps the term unit processes is more appropriate here. The important types of unit processes common to the manufacture of integrated circuits are listed in Table I. Not included are the unit processes used to manufacture masks, or the unit processes used to produce some of the materials necessary in the manufacturing process, such as the important unit processes of filtration of air and water purification. The unit processes listed may be subdivided. For example, in the unit process, "implantation", either diffusion and ion implantation can be used. Under the term, "dry etching", we can find unit processes such as gas etching, plasma etching, and reactive ion milling. Further, some overlap of the categories listed in Table 1 does exist.

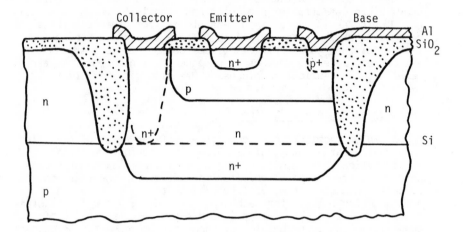

Figure 1. Example of a bipolar npn transistor. The figure shows a cross section of the silicon wafer. Symbols are: AL, aluminum; Si, silicon; SiO_2, silicon dioxide; and n and p, the doped regions of the silicon.

Table I. Categories of Unit Processes

1. Single Crystal Processess	2. Imaging Processes
a) Silica purification b) crystallization from melt c) melt doping d) zone refining e) slicing and polishing	a) coating b) baking c) exposure d) development
3. Etching and Cleaning Processes	4. Deposition and Growth Processes
a) wet etching b) dry etching	a) oxidation b) epitaxy
5. Assembly and Packaging	c) implantation d) evaporation
a) testing b) scribing c) dicing d) lead attachment e) encapsulation	e) sputtering f) lift-off g) plasma deposition h) chemical vapor deposition
6. Miscellaneous	
a) Laser-annealing	

It is instructive to consider an example of the manufacture of a typical component such as the MOS (metal-oxidide-semiconductor) transistor. Figure 2 gives a broad outline of how a p-channel MOS transistor is manufactured. Starting with a smooth and clean wafer of n-type silicon, the surface is first oxidized by exposing the wafer to oxygen gas and/or water vapor at high temperatures (~1200°C). Usually a layer of silicon dioxide approximately 1 μm thick is grown. Another process that can be used is chemical vapor deposition in which silicon dioxide is deposited from a reactive vapor phase. This latter process can be operated at a considerably lower temperature than that required for thermal oxidation and is often preferred. Silicon oxidation phenomena are well-understood (2,3) and reaction conditions can be easily designed to obtain the appropriate silicon dioxide thickness.

After the oxide layer has been grown, a thin layer of photoresist is applied to the wafer's surface. The photoresist is a radiation-sensitive polymeric solution. It contains a complex mixture of photo-sensitive molecules and solvents. The layer of photoresist must be very thin, of the order of two microns, and spin coating is a process that is used to obtain a uniform layer

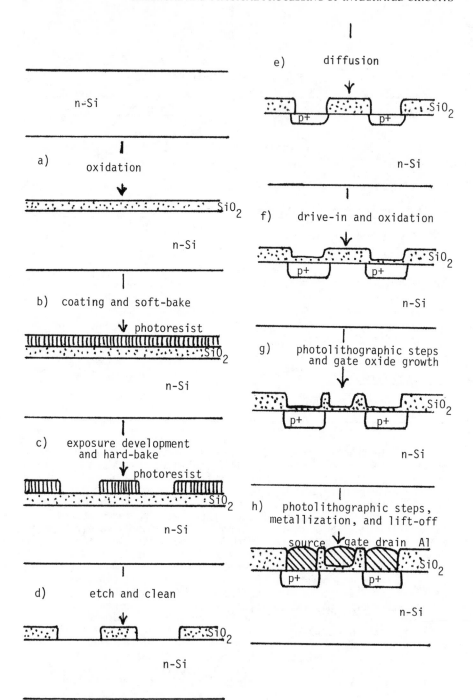

Figure 2.   Schematic diagram of the fabrication of a p-channel MOS transistor.

of photoresist on the surface. In this coating process a controlled volume of photoresist is placed on the center of the
wafer after which the wafer is spun at high speed (3000-8000 rpm)
in order to spread and thin the liquid photoresist uniformly on
the wafer. The fluid mechanics of the spin coating process is
complex and is not fully understood because of several factors:
the speed of the rotating wafer is ramped; the photoresist solution may be non-Newtonian; evaporation of solvent takes place
simultaneously with the momentum transfer process. A review
article on the problem has been given (4) and mathematical models
have appeared in the literature (5).

After spin coating the wafer is baked to remove excess solvent
from the photoresist. The baking conditions are mild (~ 90°C) and
the process is, therefore, called soft-baking. The drying of the
photoresist improves its photosensitivity to radiation, and increases adherence to the thin film layer. Next, the photoresist
is selectively exposed using an image pattern transfer technique.
For example, in proximity printing, a mask is placed very near to
the photoresist and exposed with ultra violet light through the
non-opaque areas. There are many other exposure techniques such
as optical projection, electron beam writing, and x-ray exposure.
After exposure, the development process depends on the type of
photoresist (6). For negative photoresist, the exposed areas have
undergone polymerization. Unexposed areas can be dissolved in the
developer solution. Although some components are similar, positive resists are different from negative resists in the chemistry
of the resist and developer, and the resist response to exposure.
For positive photoresist, the exposed areas become soluble in the
developer solution. After development, the pattern is cleaned by
a spray rinse and then the wafer is hard-baked. The baking temperatures can be around 200°C causing the photoresist to harden
and adhere firmly to the surface of the substrate. The lithographic steps of spin coating the photoresist, soft-baking,
exposure, and development, are always necessary as the first step
in transferring the circuit pattern from the mask to the wafer.
Insufficient information on these processes is available in the
open literature because the composition of photoresists are
proprietary. Photoresist manufacturers usually supply guidelines
to set processing conditions (7). After development, the silicon
dioxide that is not covered by photoresist can now be etched away
by either a wet etching or dry etching process. In wet etching, a
reactive solution causes dissolution of the unprotected silicon
dioxide layer. Wet etching involves a moving-boundary phenomenon,
and in recent years new mathematical models for this type of
problem have given a better insight into the physics of the process (8). In dry etching a plasma gas reacts with the silicon
dioxide and the products are vaporized at vacuum conditions. Dry
etching with plasmas is used extensively in processing, but considerable research is still in progress to understand the detailed
chemistry and process parameters (9).

The etching process results in removal of the silicon dioxide in the selected areas defined, for example, by the clear or opaque areas in the mask outlining the circuit pattern, so that the silicon is exposed.

After removal of the photoresist by a cleaning or stripping process, the exposed areas of the silicon can be doped with phosboron or other dopants to change the semicondutor from n-type to p-type in the openings of silicon dioxide (for the example shown in Figure 2). High temperature diffusion or ion-implantation can be used as the unit processes to dope the surface areas. Diffusion of impurities into solids is well-understood (10). Ion implantation is rapidly becoming a mature unit process (11). The advantage of this process is that it is carried out at considerably lower temperatures than the diffusion process although some thermal annealing may be required. Further drive-in of the purity and oxidation of the silicon can be achieved by heating the wafer in the presence of oxygen gas and/or water vapor.

Additional photolithographic steps similar to those discussed earlier are necessary to reopen the silicon dioxide covering source and drain areas above the p-type regions and to create a very thin silicon-dioxide layer below the gate area. The next step involves deposition of a metal such as aluminum to form the contacts for the source, gate and drain. Finally, an interconnection pattern is defined to connect the transistor with other electrical components on the surface. The transistors are now finished and the wafers proceed to other processing operations.

Process Flow-Sheet

The process described in the previous section can be shown as a flow-sheet as depicted in Figure 3. The figure shows the sequential nature of each of the unit processes with the repetition of the photolithographic or imaging steps involving coating, baking, exposure and development. Flow-sheets for other devices would be quite similar to Figure 3 in that similar unit processes would be employed. The particular sequence of operations may differ. Further, in any manufacturing process there is usually a choice of several unit processes that can be used to produce a particular product. For example, etching can be carried out by a wet etching or dry etching process. As far as deposition and growth processes are concerned, the unit processes in epitaxy, plasma deposition and chemical vapor deposition are extensively used to deposit other layers such as silicon nitride and polysilicon.

The process flow-sheet does not show the unit processes in preparing the wafer, nor the final assembly and packaging processes. With higher and higher density chips, the assembly and packaging may prove to be a bottleneck. The choice of a particular unit process depends on a variety of factors including experience in using the process and economics.

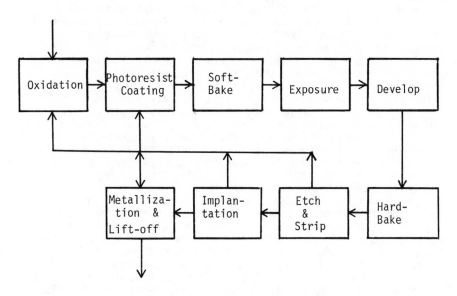

Figure 3. Process flow-sheet for the MOS transistor depicted in
Figure 2. Direction of arrows allows for the repetition of
several lithographic processes.

Fundamentals of Unit Processes

An understanding of the fundamental physical and chemical processes occurring in each unit process is essential in order that
process engineers can control the total process to insure high
yields and consistent quality of the product. In the unit processes, the basic fundamentals of transport phenomena, chemical
reaction, adsorption, phase transition, thermodynamics, etc. must
be understood in order to be able to predict what will happen to
the product if process conditions or materials are changed. For
example, the soft-baking of negative photoresist is a process of
coupled mass and heat transfer occurring in a solid in which
excess solvent must be removed. As the solvent evaporates, the
process may change from heat transfer control to mass transfer
control due to changes in the physical properties of the film.
Diffusion of solvent in polymer films is a complex phenomenon
which is not yet fully understood. In fact, the fundamental chemical and physical phenomena in many of the unit processes
currently used in the manufacture of integrated circuits are still
under intensive investigation by engineers, physicist, chemists
and material scientists.

   Besides an understanding of the basic fundamental phenomena
occurring in unit processes, the process engineer needs a solid
back-ground in material science, solid state physics and device
physics. Process materials for device manufacture are becoming
more exotic with the use of III-V and II-VI compound semiconductors. Slight changes in process conditions can change the
electrical components' characteristics drastically for such
materials. The consequence of all of this is that the engineer or
scientist working in the fabrication of integrated circuits will
either have a multidisciplinary background or will work in an
interdisciplinary group of professionals. Since some of the unit
processes are not yet fully exploited considerable research is
required to further increase our knowledge about the fundamental
phenomena occurring in such processes. Since the fabrication of
thin film devices has become of interdisciplinary nature, the
question arises on how research in this area can be most effectively communicated to and between professionals with different
educational backgrounds? Are there unifying structures by which
the knowledge on the fundamentals of unit processes can be
transmitted to the different disciplines in the best possible way?

Conclusion

The fabrication of integrated circuits can be divided in a series
of individual unit processes utilized in the manufacturing process. To control these unit processes the fundamental physical
and chemical phenomena need to be fully understood. The present
technology for making integrated circuits is versatile, but much
remains to be done to fully exploit some of the unit processes and
to develop new processes to produce cheaper and electronically
superior chips for future applications.

Literature Cited

1.   Foust A.S.; Wenzel, L.A.; Clump, C.W.; Maus, L.; Anderson, L.B. "Principles of Unit Operations", p.4, John Wiley and Sons, New York, 1960.
2.   Colglaser, R.A. "Microelectronics: Processing and Device Design", John Wiley and Sons, New York, 1980.
3.   Ghandi, S.K. "VLSI Fabrication Principles. Silicon and Gallium Arsenide", Wiley-Interscience, New York, 1983.
4.   Washo, B.D. IBM J. Res. Develop., 1977, 21, 190.
5.   Flack, W.W.; Soong, D.S.; Bell, A.T.; Hess, D.W. J. Appl. Phys. 1984, 56, 1199.
6.   DeForest, W.S., "Photoresist: Materials and Processes", McGraw-Hill Book Co., New York, 1975.
7.   Elliot, D.J. "Integrated Circuit Fabrication Technology", McGraw-Hill Book Co., New York, 1982.
8.   Kuiken, H.K. Proc. R. Soc. Lond., 1984, A392, 199.
9.   Flamm, D.L.; Donnelly, V.M.; Plasma Chem. and Plasma Process., 1981, 1, 317.
10. Grove, A.S.; Roder, A.; Sah, C.T. J. Appl. Phys., 1965, 36, 802.
11. Stone, J.L.; Plunkett, J.C. In "Impurity Processes in Silicon", Wang, F.F.Y. Ed.; North-Holland Publ. Co.: Amsterdam, 1981; Ch.2.

RECEIVED June 27, 1985

Conservation Laws for Mass, Momentum, and Energy
Application to Semiconductor Devices and Technology

Bill Baerg

Intel Corporation, Santa Clara, CA 95051

This paper describes the application of three
fundamental conservation laws for continuous media,
to three aspects of semiconductor technology: device
modeling, electromigration, and laser annealing.

The response of a semiconducting material to an applied electric
field can be described in terms of the behavior of the conduction
and valence electrons that exist within it. In a nondegenerate
semiconductor ($\underline{1}$), these electrons may be considered as a classical
ideal gas mixture, of electrons (in the conduction band) and holes
(in the valence band). The goal of this paper is to show that the
methods of continuum mechanics, as developed by C. Truesdell ($\underline{2}$),
can provide a useful description of the transport phenomena which
are observed.

First, we will briefly review the general conservation
equations for mass and momentum, for continuous media. Then we
will use these equations to describe the transport of electrons and
holes in a semiconductor. The results will correspond to those
which are used in device modeling, such as in the SEDAN ($\underline{3}$) and
MINIMOS ($\underline{4}$) programs, and will demonstrate the role of momentum
conservation.

Next, we will apply the general equations to the problem of
electromigration in a metal. In this case, the components of the
mixture will be conduction electrons and metallic ions. The motion
of the metallic ions in the interconnect constitutes a failure
mechanism, which has received renewed interest as production
circuit dimensions approach one micron ($\underline{5}$). In this work, the ions
are considered to constitute a linearly elastic material ($\underline{2}$). The
results will be compared to previous work by approximations, and a
set of equations with initial and boundary conditions, which could
be solved numerically, will be presented for the first time.

In addition to mass and momentum, energy is also conserved.
The basic equations will be reviewed, and applied to laser
annealing. Laser annealing is, among other things, a promising
method of recrystallizing the damaged surface layers of a

semiconductor following ion implantation (6). Instead of to an
applied electric field in the usual sense, the electron and hole
transport will be in response to incident optical radiation.

Conservation of Mass

Consider a gas mixture of several components, labeled with the
subscript A : A = 1, 2, ... Each component is made up of
particles with mass m_A and has a density of n_A particles per
unit volume, where n_A is a mathematically continuous,
differentiable function of space and time. Further, consider an
arbitrary volume V of the mixture, enclosed by a surface S. Then
the mass of component A is conserved if

$$\frac{\partial}{\partial t} \int_V m_A n_A dV = -\oint_S m_A n_A v_A \cdot n dS + \int_V m_A (G_A - R_A) dV \tag{1}$$

where v_A is the velocity vector of component A, and **n** is
the outer normal unit vector for S. (Vectors are identified with
boldface letters.) G_A and R_A are the generation and
recombination rates, respectively, for component A. Equation 1
states that, for component A, in unit time, the increase in the
mass in V is equal to the flux of mass from outside V, plus the net
generation of mass within V. The surface integral has a minus sign
because **n** is directed outward from V. Since m_A is constant,
and appears in all three integrals, it cancels out. Also, we can
apply the divergence theorem, converting the surface integral to a
volume integral:

$$\int_V \left[\frac{\partial n_A}{\partial t} + \nabla \cdot n_A v_A - (G_A - R_A) \right] dV = 0 \tag{2}$$

Since V is arbitrary, the integral must vanish, and

$$\frac{\partial n_A}{\partial t} + \nabla \cdot n_A v_A = G_A - R_A \tag{3}$$

which is the continuity equation for component A. When the gas
mixture consists of particles with charge q_A, we may define an
electrical current density, **j**, so that

$$j_A = q_A n_A v_A \tag{4}$$

Then the continuity Equation 3 becomes the conservation of charge:

$$\frac{\partial}{\partial t} (q_A n_A) + \nabla \cdot j_A = q_A (G_A - R_A) \tag{5}$$

For a binary gas mixture of electrons and holes, we define n and p
as the electron and hole number densities, respectively, and also
use the same letters as subscripts to represent the electron and
hole components. Then we can write the charge Equation 5 for holes
of charge +q, and electrons of charge -q, respectively:

$$\frac{\partial}{\partial t} (qp) + \nabla \cdot \mathbf{j}_p = q(G_p - R_p) \qquad (6)$$

and

$$\frac{\partial}{\partial t} (-qn) + \nabla \cdot \mathbf{j}_n = -q(G_n - R_n) \qquad (7)$$

Now let's consider the total charge density in a semiconductor,

$$\rho_Q = q \, (p - n + N_D - N_A + (1-f) \, n_T) \qquad (8)$$

where N_D and N_A are the dopant donor and acceptor concentrations, respectively, n_T is the trap density, and f is the fraction of traps occupied by electrons. Here we have in mind the Shockley-Read-Hall recombination theory (7), and the traps are sites within the semiconductor, which have energy levels within the band gap. These sites may be due to crystal dislocations, impurity atoms or surface defects. Each site may be occupied by an electron, making the trap neutral, or a hole, making it positively charged. The trap density n_T, as well as N_D and N_A, is constant in time, but this is not generally true for f: for example, in a time-dependent electric field, an MOS capacitor with p-type Silicon has surface traps which may be filled with holes (accumulation) or electrons (inversion) depending on the electric field. Therefore, taking the time derivative of Equation 8 gives

$$\frac{\partial \rho_Q}{\partial t} = q \frac{\partial}{\partial t} \left(p - n + n_T (1-f) \right) \qquad (9)$$

Now we can add Equations 6 and 7, so that

$$q \frac{\partial}{\partial t} (p - n) + \nabla \cdot \mathbf{j} = -q(U_p - U_p) \qquad (10)$$

where

$$\mathbf{j} = \mathbf{j}_p + \mathbf{j}_n \qquad (11)$$

is the total current density, and

$$U_p = -(G_p - R_p)$$

and

$$U_n = -(G_n - R_n) \qquad (12)$$

are the net recombination rates for holes and electrons, respectively. Adding Equations 9 and 10,

$$\frac{\partial \rho_Q}{\partial t} + \nabla \cdot \mathbf{j} = q n_T \frac{\partial (1-f)}{\partial t} - q \, (U_p - U_n) \qquad (13)$$

Since there is no creation of net charge,

$$\frac{\partial \rho_Q}{\partial t} + \nabla \cdot \mathbf{j} = 0 \qquad (14)$$

and therefore,

$$-n_T \frac{\partial f}{\partial t} = U_p - U_n \qquad (15)$$

The Shockley-Read-Hall theory assumes a steady-state condition, so that, $U_p = U_n$, since $\frac{\partial f}{\partial t} = 0$. A recombination theory for the more general case does not yet exist.

Conservation of Momentum

In order to derive the momentum equations, we must consider an integral over a volume which is not fixed in space. Rather, the integral is taken over the set of particles which define an arbitrary volume of the gas mixture at some instant in time. As these particles move in space, so does the volume of integration. Thus the volume of integration changes with time. The force F_A acting on component A in a volume $V(t)$ of the gas mixture, enclosed by a surface $S(t)$, is

$$F = \oint_S t_A ds + \int_V f_A dV + \int_V p_A dV \qquad (16)$$

where f_A is the external or applied force density exerted on component A, p_A is the force density exerted on A by the other components, and t_A is the stress vector for component A(2).

The stress vector represents the force of the material outside of $V(t)$ acting on the material inside $V(t)$. The assumption, based on the short range of the inter-particle forces, is that this force which the material exerts on itself acts entirely on the surface $S(t)$. Truesdell calls this stress principle the defining concept of continuum mechanics. The stress vector depends not only on time and space, but also on the surface orientation. However, according to Cauchy's fundamental theorem,

$$t_A = T_A \cdot n \qquad (17)$$

where n is the unit normal, as before, and T_A is the stress tensor. The stress tensor depends only on time and space, and is not a physical quantity, but a concept. A particular material is defined in this regard by a constitutive equation, relating the stress tensor to its motion.

The subscript A means that the stress tensor represents only the force of component A acting on itself. In addition, there is the force exerted on component A by the other components in the system. We call this force density p_A.

The law of conservation of momentum states that the total force acting on the material is equal to the rate of change of its momentum:

$$F_A = \oint_{S(t)} T_A \cdot n ds + \int_{V(t)} (f_A + p_A) dV = \frac{d}{dt} \int_{V(t)} \rho_A v_A dV \qquad (18)$$

The last term is proportional to the mass of a particle of · component A, whereas the applied force density for a coulomb force, which we consider below, is proportional to the charge of a particle of component A. The small mass-to-charge ratio of an electron has the effect that this acceleration term is negligible under ordinary conditions, as will be shown at the end of this section, and we make this assumption throughout this paper.

As a result, the momentum conservation law reduces to the statement that the total force F_A on a volume moving with the material must vanish. Applying the divergence theorem,

$$F_A = \int_{V(t)} (\nabla \cdot T_A + f_A + p_A) dV = 0 \qquad (19)$$

Since the volume is arbitrary, and the integrand is assumed continuous,

$$p_A + f_A + \nabla \cdot T_A = 0 \tag{20}$$

The total applied force acting on the mixture is,

$$f = \sum_A f_A \tag{21}$$

and the total stress is

$$T = \sum_A T_A \tag{22}$$

where we have neglected a term on the same order of smallness as the acceleration term in Equation 18.

Then summing Equation 20 over all components in the mixture,

$$\sum_A p_A + f + \nabla \cdot T = 0 \tag{23}$$

But the total force for the mixture as a whole must also vanish, so

$$f + \nabla \cdot T = 0 \tag{24}$$

The last equation is valid for any material, regardless of the forces which the components of the mixture may exert on each other (neglecting accelerations). Therefore

$$\sum_A p_A = 0 \tag{25}$$

Equation 20 is our momentum, or force balance equation which we may now apply to semiconductors. To do so, we will need constitutive equations for the p_A, subject to condition Equation 25, and for the stress tensors T_A.

Device Modeling Equations

In order to model electron and hole transport in a semiconductor device, such as a transistor, we need solutions to a set of equations for the concentrations n and p, the electric current densities j_n and j_p, and for the electrostatic potential ϕ, defined by $E = -\nabla\phi$. Since there are nine unknowns (counting a vector as three, for each component), we must have nine equations. Two of them have already been derived: Equations 6 and 7, for the conservation of charge. Another is Poisson's equation for the electric field:

$$\nabla \cdot D = \rho_Q \tag{26}$$

with ρ_Q given by Equation (8). D is the electric displacement which, for a simple material, is given by

$$D = \varepsilon E \tag{27}$$

where ε is the dielectric constant of the material. Therefore, Poisson's equation becomes

$$\varepsilon \nabla^2 \phi = -\rho_Q \tag{28}$$

The remaining six equations are the (vector) momentum Equations 20, for electrons and holes.

The coulomb force density on component A is

$$f_A = q_A n_A E \tag{29}$$

The additional (non-coulomb) force density on component A due to the other components must include the interaction of the fixed ions and atoms in the semiconductor crystal. This force is modeled as a drag force, analogous to the drag of a fluid moving past a solid sphere, and is proportional to the relative velocity of each component. The drag force of the electrons on the holes, and vice versa, is neglected, which is an assumption of diluteness of the gas mixture relative to the density of the fixed atoms. The drag of the crystal on the electrons is given by

$$p_n = j_n/\mu_n \tag{30}$$

where μ_n is the electron mobility. The effects of the atoms, or lattice scattering, the dopant ions, or ionized impurity scattering, and defects (which are particularly important at an interface), are all lumped into this definition of mobility. Even when the diluteness assumption is invalid (in low-doped, high injection regions), the mobility is empirically adjusted to fit the data.

Similarly, for holes,

$$p_p = -j_p/\mu_p \tag{31}$$

where μ_p is the hole mobility. The difference in sign is due to the positive and negative charges of electrons and holes. The hole current and velocity vectors have the same direction, or sign, and the hole drag force has the opposite direction.

The condition Equation 25, which states that $\Sigma P_A = 0$, is not of any value here since it merely tells us that the drag force of the electrons plus holes on the crystal is equal to the drag force of the crystal on the electrons and holes. It is not true, of course, that $p_n + p_p = 0$.

Finally, we assume the drag force is proportional to j_A, rather than just v_A, because it is more plausible to allow the density and velocity to vary inversely without affecting the drag, and $j_A = q_A n_A v_A$ is a convenient variable.

The divergence of the stress tensor, T_A, represents the force which component A exerts on itself. For a perfect fluid,

$$T_A = -p_A 1 \tag{32}$$

where 1 is a unit tensor (with components δ_{ij}), and P_A is the partial pressure of component A. For a mixture of ideal gases,

$$p_A = n_A k T \tag{33}$$

where k is the Boltzmann constant and T is the temperature (not to be confused with the total stress T, which is used only once, in Equation 22. So, at constant temperature,

$$\nabla.T_A = -\nabla p_A = -kT\nabla n_A \tag{34}$$

Now we can substitute Equations 34, 29, and 30 or 31 in Equation 20.

$$\pm j_A/\mu_A + q_A n_A E - kT\nabla n_A = 0 \tag{35}$$

For electrons and holes, respectively, Equation 35 is

$$j_n = \mu_n q n E + \mu_n kT\nabla_n$$
$$j_p = \mu_p q p E - \mu_p kT\nabla_p \tag{36}$$

These equations complete the set required for device modeling, such as SEDAN or MINIMOS. We repeat the others below:

$$\frac{\partial n}{\partial t} - \frac{\nabla.j}{q}_n = -U_n \tag{6}$$

$$\frac{\partial p}{\partial t} + \frac{\nabla.j}{q}_p = U_p \tag{7}$$

and

$$\varepsilon\nabla^2\phi = -q\ (p - n + N_D - N_A + (1-f)n_T) \tag{28}$$

where we have used Equations 8 and 12. The unknowns are n, p, j_n, j_p, and ϕ, with $E = -\nabla\phi$. In addition, the quantities μ_n, μ_p U_p, U_n, and f must be given in terms of the unknowns, and N_D, N_A and n_T must be given. Note that the electric field, through the potential ϕ, is related to the net charge density through Poisson's equation, even though the field is externally applied (to the terminals of the device).

To summarize this section, we have seen that device modeling can be performed with a set of equations for mass and momentum conservation, plus Poisson's equation for the electric field. We have obtained the conventional equations from the general laws for continuous media by neglecting accelerations, by treating the electrons and holes as an ideal gas mixture, and by introducing the concept of the drag force (originated by Truesdell for general mixtures) between the electrons or holes, and the crystal.

Acceleration

In Equation 18, the acceleration term would be significant if it were comparable in magnitude to the coulomb term, or if

$$qnE \simeq mn\frac{dv}{dt} \tag{37}$$

We consider an MOS transistor with channel length ℓ, and ask at what value of ℓ does Equation 37 hold? If the transit time for an electron to move the distance ℓ is τ, then

$$\frac{dv}{dt} \simeq \frac{v}{\tau} \simeq \frac{\ell}{\tau}2 \tag{38}$$

But, we also have

$$y \simeq \mu E \tag{39}$$

where μ is the electron mobility in the channel, so

$$\tau \simeq \frac{\ell}{v} \simeq \frac{\ell}{\mu E} \tag{40}$$

Therefore,

$$\frac{dy}{dt} \simeq \frac{(\mu E)^2}{\ell} \tag{41}$$

Equation 41 gives the electron acceleration in the channel in terms of the channel length, mobility, and electric field. Substituting in Equation 37, and using

$$E \simeq y/\ell \tag{42}$$

where V is the potential drop across ℓ, we obtain

$$\ell \simeq \mu y \sqrt{\frac{m}{qy}} \tag{43}$$

Substituting numerical values,

$\mu = 500$ cm^2 volt^{-1} sec^{-1},

$V = 5$ volt,

$m = 9.1 \times 10^{-28}$gm

and $qV = 5eV = 8.0 \times 10^{-12}$erg, we find $\ell = 0.27\mu$m.

So, for this MOS example we find that acceleration will become important when the channel length approaches a quarter-micron. However, from Equation 43, the channel length at which acceleration becomes important is proportional to the mobility, so a high-mobility material would have a longer ℓ. The effect of the acceleration will be to reduce the effective electron mobility in the crystal.

Electromigration

Electromigration is the displacement of atoms in a conductor due to an electric current. A metal, such as aluminum, consists of positively charged ions, and Z conduction electrons per ion (Z = 3 for aluminum). To simplify the discussions, we will consider an idealized homogenous conductor, rather than the more realistic polycrystalline description. Our mixture is then made of two components, an electron gas and an ionic body, which we will describe as a linearly elastic solid.

There are no generation or recombination mechanisms in a metal, so the continuity Equation 3 becomes

$$\frac{\partial n_e}{\partial t} + \nabla \cdot n_e v_e = 0 \tag{44}$$

for electrons, and

$$\frac{\partial n_i}{\partial t} + \nabla \cdot n_i v_i = 0 \tag{45}$$

for the ions.

In the momentum Equation 20, we again neglect accelerations, and use the coulomb force Equation 29 for the applied force on ions and electrons, with the electric field given by Poisson's Equation 28. That is,

$$f_i = Zqn_iE \tag{46}$$

$$f_e = qn_eE \tag{47}$$

and

$$\varepsilon\nabla.E = q(Zn_i-n_e) \tag{48}$$

since the ionic charge is Zq. The dielectric constant of the metal is ε.

The force densities of the electrons acting on the ions (the electron "wind") and of the ions acting on the electrons must balance, by Equation 25, since there are only two components in the mixture (in the semiconductor model there were three : electrons, holes and fixed ions or atoms). Hence, as in Equation 30, we have

$$p_e = j_e/\mu_e \tag{49}$$

for the electrons, where μ_e is the electron mobility in the metal, and since by Equation 25,

$$p_e + p_i = 0 \tag{50}$$

$$p_i = -j_e/\mu_e \tag{51}$$

we are neglecting v_i compared to v_e in the model for the drag force, which is more precisely proportional to (v_e -v_i). Note that there is no ion mobility coefficient involved. However, we are not neglecting v_i elsewhere, such as in the continuity Equation 45, since v_i describes the electromigration. Now we can substitute these results in Equation 20, for the ions

$$j_e/\mu_e = Zqn_iE + \nabla.T_i \tag{52}$$

and for the electrons,

$$j_e/\mu_e = qn_eE-\nabla.T_e \tag{53}$$

The stress tensors are T_i and T_e for the ions and conduction electrons, respectively. Before we discuss constitutive equations for the stress, we note by subtracting Equation 53 from Equation 52 that

$$\nabla.T_i = -q(Zn_i-n_e)E-\nabla.T_e \tag{54}$$

This equation strongly suggests space charge plays an important role in electromigration, since if the divergence of the ionic stress tensor vanishes, there isn't any electromigration. So, we will not want to assume that $Zn_i = n_e$ in this problem.

The nine Equations 44, 45, 48, 52, and 53 form a complete set, which we repeat below:

$$\frac{\partial n_e}{\partial t} + \nabla.n_e v_e = 0 \tag{44}$$

$$\frac{\partial n_i}{\partial t} + \nabla.n_i v_i = 0 \tag{45}$$

$$j_e/\mu_e = qn_i E + \nabla.T_i \tag{52}$$

$$j_e/\mu_e = qn_e E - \nabla.T_e \tag{53}$$

$$\varepsilon\nabla.E = q(Zn_i - n_e) \tag{48}$$

where $E = -\nabla\phi$ as before. The nine unknowns are n_i, n_e, v_i, v_e and ϕ. The coefficients μ_e and ε must be given, as well as constitutive equations for T_i and T_e in terms of the unknowns. Finally, boundary and initial conditions must be given.

The stress tensor models, especially the one for the ions, play key roles in this problem. Comparison of these equations with the ones for a semiconductor shows that the differences are in the absence of source terms in the continuity equations, the absence of impurity ions or traps, and in the stress tensors.

We start with the simpler model, for the electron gas. In a metal, the electron density is so high that the Pauli exclusion principle must be taken into account, i.e., there can be only one electron in each quantum state. The result is that the electrons obey the Fermi distribution rather than Boltzman's, and may be considered strongly degenerate (8). The equation of state is

$$P_e = \frac{(3\pi^2)^{2/3}}{5} \frac{\hbar^2}{m_e} n_e^{5/3} \tag{55}$$

As before, we assume the electron gas is a perfect fluid so that Equation 30 applies, and

$$\nabla.T_e = -\nabla P_e$$
$$= \frac{\hbar^2}{3m_e}(3\pi^2 n_e)^{2/3} \nabla n_e \tag{56}$$

This result is, of course, non-linear in the electron density. Since we have considered only first order or linear terms throughout this theory, it seems inconsistent to include a non-linear term here. Therefore, we will revert to the ideal gas model for the electrons, so that

$$\nabla.T_e = -kT\nabla n_e \tag{57}$$

We now turn to the stress tensor for the metallic ions, making use of the theory of elasticity: Let x_0 be the vector describing the position of a particle in the body before it is deformed, and let x be the vector to this same particle after a deformation. Then

$$u = x - x_0 \tag{58}$$

is called the displacement vector and, for small deformations, the strain tensor E has the components

$$E_{k\ell} = \frac{1}{2}\left(\frac{\partial u_k}{\partial x_\ell} + \frac{\partial u_\ell}{\partial x_k}\right)$$

(59)

A perfectly elastic body is one whose stress arises solely in response to the strain from its original state, and if the response is also linear and isotropic, the material obeys Hooke's Law (9):

$$T_i = \lambda(\text{trace } E)1 + 2GE$$

(60)

The coefficients λ and G are called Lame's constants, and are related to the Young's modulus E and Poisson's ratio ν by

$$\lambda = \frac{E\nu}{(1 + \nu)(1 - 2\nu)}$$

(61)

and

$$G = \frac{E}{2(1 + \nu)}$$

(62)

Note that by Equation 58, trace $E = E_{kk} = \nabla.u$. Then

$$\nabla.T_i = (\lambda + G)\nabla(\nabla.u) + G\nabla^2 u$$

(63)

where ∇^2 is the Laplacian.

We now have, with Equations 57 and 63, the relations for T_e and T_i in terms of the unknowns in Equations 52 and 53, because the displacement vector u and the velocity vector v_i are related by

$$v_i = \frac{\partial u}{\partial t}$$

(64)

Therefore, the unknown v_i should be replaced by u.

Let's write down the complete set of equations again:

$$\frac{\partial n_e}{\partial t} - \frac{1}{q}\nabla.j_e = 0$$

(65)

$$\frac{\partial n_i}{\partial t} + \nabla.\left(n_i\frac{\partial u}{\partial t}\right) = 0$$

(66)

$$j_e/\mu_e = Zqn_iE + (\lambda + G)\nabla(\nabla.u) + G\nabla^2 u$$

(67)

$$j_e/\mu_e = qn_eE + kT\nabla n$$

(68)

and

$$\varepsilon\nabla.E = q(Zn_i - n_e)$$

(69)

Now, we consider spatial variation in only one dimension. We have

$$\frac{\partial n_e}{\partial t} - \frac{1}{q}\frac{\partial j_e}{\partial x} = 0$$

(70)

$$\frac{\partial n_i}{\partial t} + \frac{\partial}{\partial x}\left(n_i\frac{\partial u}{\partial t}\right) = 0$$

(71)

$$j_e/\mu_e = Zqn_iE + (\lambda + 2G)\frac{\partial^2 u}{\partial x^2}$$

(72)

$$j_e/\mu_e = qn_eE + kT\,\frac{\partial n_e}{\partial x} \tag{73}$$

$$\varepsilon\frac{\partial E}{\partial x} = q(Zn_i - n_e) \tag{74}$$

where now the unknowns n_i, n_e, u, j_e, and E are all scalars.

The solution to these equations must await future work. However, we can discuss a possible set of boundary and initial conditions of interest. Imagine a block of metal, infinite in the y and z directions, sandwiched between two blocks of n^+ silicon, at $x = 0$ and $x = L$ (Fig. 1). The electron current is to the left, so the electron velocity, and metal ion displacement, are to the right. We could imagine the metal to have initially a certain uniform vacancy density, so for n_i (x, t), we would have n_i $(x, 0) = N_0$, where $N_0 < N_{max}$.

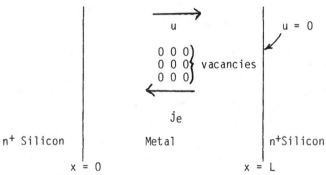

Figure 1.   Electromigration in a metal block between two blocks of n^+ silicon.

N_{max} would be the theoretical density of the "perfect" metal (i.e., without vacancies). Then we could define the time to failure, t_f, such that n_i $(0, t_f) = 0$. In other words, t_f would be the time for all the metal ions initially at $x = 0$ to migrate to vacancies somewhere to the right. Since the number of vacancies available per cm^2 would be proportional to L, we would expect that t_f would be inversely proportional to L, as was experimentally observed by Blech (10). If we imagine that the current density j_e is produced by a pulsed dc technique, so that joule heating is slight, then dissolution of silicon at the interface could be prevented. Then $u(L,t) = 0$ would be an appropriate boundary condition.

Since the equilibrium electron density in the n_+ silicon is 3 or 4 orders of magnitude less than in the metal ($n_e = 1.8 \times 10^{23}$ cm^{-3} for aluminum), we would expect a large gradient in n_e at $x = 0$, as well as a space charge region, thus satisfying the requirement Equation 54 for electromigration.

Additional boundary conditions could be obtained from the requirements of continuity for j_e and $D = \varepsilon E$ at the Si/Al interfaces.

In order to compare the set of Equations 70 - 74 with previous work, consider first the following result of Blech and Herring (11):

$$qZ^*E = \frac{1}{n_i} \frac{\partial T_{xx}}{\partial x}$$

(75)

where Z^* is an average effective charge, and T_{xx} is the normal stress. If we neglect the coulomb force compared to the stress in the force balance for the ions, and vice-versa for the electrons, then

$$j_e/\mu_e = Zqn_iE + \frac{\partial T_i}{\partial x} \text{ (cancel)}$$

(76)

$$j_e/\mu_e = qn_eE + \frac{\partial T_e}{\partial x} \text{ (cancel)}$$

(77)

where T_i is the normal ion stress component. Finally, if $n_e = Zn_i$, then

$$\frac{\partial T_i}{\partial x} = qn_eE = qZn_iE$$

(78)

or

$$qZE = \frac{1}{n_i} \frac{\partial T_i}{\partial x}$$

(79)

which compares to Equation 75 except for the Z instead of Z^*. Apparently the Z^* is based on the concept that the ions do not behave as point charges with a charge density Zn_i, as they are assumed to do in this paper. However, the assumptions required to obtain Equation 79, namely the neglect of the electron density gradient and space charge, do not seem justifiable in the light of Equation 54.

Finally, we compare with the following result of Black (12):

$$\frac{1}{MTTF} = A \exp(-\Delta E_a/kT)j_e^2$$

(80)

where MTTF is the median-time to failure of the line, A is a constant which includes the volume resistivity of the metal, the electron free-time between collisions, and other factors, and ΔE_a is an activation energy, equal to about 0.5eV for aluminum.

Consider a semi-infinite layer of thickness δ. Initially, there are n_i atoms/cm^2 in this layer. The removal rate of atoms from the layer is $\delta \frac{\partial n_i}{\partial t}$ per cm^2 per sec., so if the time for all the atoms in the layer to be removed is approximately the MTTF, then

$$MTTF = - \frac{\delta n_i}{\delta \frac{\partial n_i}{\partial t}}$$

(81)

$$\frac{1}{MTTF} = - \frac{1}{n_i} \frac{\partial n_i}{\partial t}$$

(82)

From the continuity Equation 71, within δ, we have approximately,

$$\frac{1}{n_i} \frac{\partial n_i}{\partial t} + \frac{\partial}{\partial x} \frac{\partial u}{\partial t} = 0$$

(83)

so

$$\frac{1}{MTTF} = \frac{\partial}{\partial t} \frac{\partial u}{\partial x}$$

(84)

where we have exchanged the order of differentiation. Now we can obtain an expression for $\partial u/\partial x$ from Equations 72 - 74, as follows: First, eliminating j_e/μ_e from Equations 72 and 73,

$$(Zn_i-n_e)qE + (\lambda + 2G)\frac{\partial^2 u}{\partial x^2} - kT\frac{\partial n_e}{\partial x} = 0 \qquad (85)$$

Putting Equation 74 into the first term gives

$$\frac{\varepsilon}{2}\frac{\partial(E^2)}{\partial x} + (\lambda + 2G)\frac{\partial^2 u}{\partial x^2} - kT\frac{\partial n_e}{\partial x} = 0 \qquad (86)$$

Now, integrating,

$$\frac{\varepsilon}{2} E^2 + (\lambda + 2G)\frac{\partial u}{\partial x} - kTn_e = 0 \qquad (87)$$

or

$$\frac{\partial u}{\partial x} = \frac{1}{\lambda + 2G}\left(\frac{-\varepsilon}{2} E^2 + kTn_e\right) + \text{constant} \qquad (88)$$

Consider only the first term on the right, and take the time dertivative:

$$\frac{\partial}{\partial t}\frac{\partial u}{\partial x} = -\frac{1}{2}\frac{\partial}{\partial t}\left(\frac{\varepsilon E^2}{\lambda + 2G}\right) \qquad (89)$$

If we assume the variations in ε and E are small, we can take them outside the time derivative, and we have

$$\frac{1}{\text{MTTF}} = \frac{\varepsilon E^2}{2(\lambda + 2G)^2}\frac{\partial}{\partial t}(\lambda + 2G) \qquad (90)$$

The time derivative can be replaced with the Maxwellian relaxation time τ, which is the time during which the stresses in the metal are damped after the applied forces are removed (11).
Then Equation 90 becomes

$$\frac{1}{\text{MTTF}} = \frac{\varepsilon E^2}{2\tau(\lambda + 2G)} \qquad (91)$$

Finally, if we neglect the electron concentration gradient in Equation 73, $j_e = q\mu_e n_e E$, and substitution in Equation 91 gives

$$\frac{1}{\text{MTTF}} = \frac{\varepsilon}{2q^2 \cdot \mu_e^2 n_e^2 \tau(\lambda + 2G)} j_e^2 \qquad (92)$$

This is the equation we wish to compare with Equation 80. Equations 92 amd 80 are consistent if

$$A \exp(-\Delta E_a/kT) = \frac{\varepsilon}{2q^2 \mu_e^2 n_e^2 \tau(\lambda + 2G)} \qquad (93)$$

The exact temperature dependence of the Right-Hand-Side has not yet been determined. The bulk modulus K is related to Lame's constants by

$$K = \lambda + \frac{2}{3} G \qquad (94)$$

Also, the viscosity is related to the bulk modulus by

$$\eta \simeq K\tau \qquad (95)$$

for a highly viscous material (11). Therefore we can say, roughly, that the R.H.S. is inversely proportional to the square of the electron mobility, and to the viscosity of the metal. Both of these quantities would decrease with increasing temperature, so the MTTF would also decrease, as it should.

Conservation of Energy/Laser Annealing

The equation of energy conservation takes a very simple form if we do not include the effects of the motion of the body. It can be applied to the atoms of a crystal, for example: We have

$$n_a \frac{\partial E_a}{\partial t} = -\nabla \cdot h_a + n_a Q_a + n_a E_a \qquad (96)$$

where E_a is the internal energy per atom, h_a is the heat flux from surrounding parts of the body,

$$h_a = -\kappa_a \nabla T \qquad (97)$$

Q_a is the energy per atom supplied by an external source to the body, and E_a is the energy per atom supplied by the other components in the body.

We want to apply this equation to a semiconductor crystal which is being heated by incident radiation from a laser. We make the problem one-dimensional, and consider incident radiation of intensity I of which a fraction R is reflected, and which has an absorption coefficient α in the material:

$$I = I_0(1-R)e^{-\alpha x} \qquad (98)$$

Then the amount of energy absorbed by the crystal is

$$\frac{-dI}{dx} = \alpha I_0(1-R)e^{-\alpha x} \qquad (99)$$

and the number of photons of energy $\hbar\omega$ per sec. which are absorbed (annihilated) is

$$\frac{\alpha I_0}{\hbar\omega}(1-R)e^{-\alpha x} \qquad (100)$$

Each annihilated photon is assumed to create an electron-hole pair in the crystal. The electrons and holes are assumed to form an ideal gas mixture at the same temperature as the crystal. The kinetic energy of a monatomic ideal gas is $\frac{3}{2}kT$. If the energy of the incident photon exceeds the bandgap, i.e., if $\hbar\omega > E_g$, then $\hbar\omega -$ E_g is the combined kinetic energy of the generated electron plus hole. But, on the average, only $\frac{3}{2}kT$ of this is needed for each carrier type to reach the common temperature T with the atoms in the lattice. Therefore, the average combined excess energy is $\hbar\omega -$ E_g - 3kT for each electron plus each hole.

In addition, when the electron and hole recombine, all of their energy, E_g + 3kT, is assumed to be transferred non-radiatively to a third carrier via Auger recombination. But this also becomes excess energy, i.e., more than is needed to maintain the temperature T, and is, therefore, transferred to the lattice atoms. The Auger recombination rate is γn_e^3, where γ is the Auger coefficient.

Now we can write for E_a, the energy supplied to the atoms by the carriers,

$$E_a = \hbar\omega - (E_g + 3kT) \frac{\alpha I_0}{\hbar\omega}(1-R)e^{-\alpha x} + (E_g + 3kT)\gamma n_e^3 \qquad (101)$$

The internal energy Ea of a solid is $E_0 + C_aT$, where $Ca = Cp = Cv$ is the heat capacity per atom.

Therefore, $\partial Ea/\partial t = C_a \partial T/\partial t$. Finally, we can substitute Equations 97 and 101 in 96, and setting $Q_a = 0$, we find

$$\frac{\partial T}{\partial t} = \frac{1}{n_a}\frac{\partial}{\partial x}\left(\frac{K_a}{C_a}\frac{\partial T}{\partial x}\right) + \left[\hbar\omega - (Eg + 3KT)\right]\left(\frac{\alpha I_0(1-R)}{\hbar\omega C_a}\right)e^{-\alpha x} + \frac{Eg+3KT}{C_a}\frac{\gamma n_e^3}{}\quad (102)$$

This equation is the same as the one derived by Van Driel, et al, in Reference (13). We have shown that the extra terms in the heat conduction equation can be derived from Truesdell's energy equation by making use of the supply term, which describes the energy received by one component (the atoms) from other components (the carriers) in the material.

The equation for the electron or hole current used by Van Driel, et al, does not seem at first glance to be a momentum equation:

$$j_e = \mu_e n_e\left[\nabla EF - \left(\frac{E_F - E_C}{T} - \frac{5}{2}k\right)\nabla T\right] \quad (103)$$

where E_f is the quasi-Fermi level, and E_C is energy of the conduction band edge. However, we can obtain this from the momentum Equation 53, if we assume $E = 0$ (a neutral electron-hole gas mixture) and $T_e = -p_e1$. Then

$$j_e = \mu_e \nabla p_e \quad (104)$$

But from thermodynamics (10),

$$d\tilde{\mu}_e = -s_e dT + \frac{1}{n_e}\,dp_e \quad (105)$$

where $\tilde{\mu}_e$ is the electron chemical potential and s_e is the entropy per electron. However, we also know that $\tilde{\mu}_e = E_f$ in a semiconductor (1), and further, that for a monatomic ideal gas (1),

$$s_e = -\frac{E_F - E_C}{T} + \frac{5}{2}k \quad (106)$$

Equation 105 can be written with gradients,

$$\nabla p_e = n_e\nabla\tilde{\mu}_e + n_e s_e\nabla T \quad (107)$$

and substituted in Equation 104:

$$j_e = \mu_e n_e\nabla\tilde{\mu}_e + \mu_e n_e s_e\nabla T$$
$$= \mu_e n_e\left[\nabla E_F - \left(\frac{E_F - E_C}{T} - \frac{5}{2}k\right)\nabla T\right] \quad (108)$$

where we have used Equation 106 and $\tilde{\mu}_e = E_F$. Since this is Equation 103, we have shown that Equation 103 is really a momentum, or force balance, equation.

Conclusions

We have seen how continuum mechanics can be applied to semiconductor problems. In particular, we have shown how the usual current equations of device modeling are derived from momentum conservation, and how the carrier mobilities are related to momentum exchange with the lattice. We saw the role of the stress tensor, and how the carriers are treated as ideal gases.

Then we derived a set of equations describing electromigration, using Hooke's law for the stress tensor, and by approximation, compared the results with previous work. We found that the mean-time-to-failure is approximately proportional to the square of the conductivity, and to the viscosity of the metal.

Finally, we considered laser annealing, and showed that some previous work can be formulated with Truesdell's energy equation, by treating the radiant energy as generating carriers which, in turn, transfer energy to the atoms in the crystal. Also, we saw how the quasi-Fermi level is related to the stress tensor, and again confirmed that the current equation is a statement of momentum conservation.

Acknowledgment

I want to thank O. Memelink, J. Middlehoek and H. Wallinga for their hospitality at Twente University of Technology, Enschede, The Netherlands, and G. Parker and G. Moore of Intel for supporting my extended sabbatical leave.

Nomenclature

Symbol	Description	Units (cgs)
C	Heat Capacity	$erg(°K)^{-1}$
E	Electric field vector	esu
E	Strain tensor	dimensionless
E	Young's modulus	$dyne\ cm^{-1}$
E	Internal energy per particle	erg
E	Energy supplied, per particle, by other components	erg
E_F	Electron energy at the quasi-Fermi level	erg
E_c	Electron energy at the conduction band edge	erg
E_g	Band gap energy	erg
f	Applied force density	$dyne\ cm^{-3}$
G	Lame constant	$dyne\ cm^{-2}$
G	Carrier generation rate	cm^{-3}
ℏ	Planck's constant, $h/2\pi$	erg sec
h	Heat flux vector	$erg\ cm^{-2}\ sec^{-1}$
I	Radiation intensity	$erg\ cm^{-2}\ sec^{-1}$
j	Current density	$esu\ cm^{-2}$
k	Boltzman Constant	$erg\ (°K)^{-1}$
ℓ	MOS transistor channel length	cm
m	Mass	gm
n	Electron density	cm^{-3}
n	Outer normal unit vector	dimensionless
p	Hole density	cm^{-3}
p	Drag force density	$dyne\ cm^{-3}$
P	Partial pressure	$dyne\ cm^{-2}$
q	Charge of a particle (4.8×10^{10} esu for a hole)	esu
Q	External energy per particle	erg
R	Reflection coefficient	dimensionless
R	Carrier recombination rate	cm^{-3}

Symbol	Description	Units (erg)
s	Entropy per particle	erg $(°K)^{-1}$
S	Arbitrary surface	cm^2
t	Stress vector	dyne cm^{-2}
t	Time	sec
T	Stress tensor	dyne cm^{-2}
T	Temperature	$°K$
u	Displacement vector	cm
v	Velocity	cm sec^{-1}
V	Arbitrary volume	cm^3
Z	Valence, or number of conduction electrons per ion	dimensionless
α	Absorption coefficient	cm^{-1}
ϵ	Dielectric coefficient	dimensionless
λ	Lame constant	dyne cm^{-2}
μ	Carrier mobility	$cm^2 volt^{-1}sec^{-1}$
$\tilde{\mu}$	Chemical potential	erg
ν	Poisson's ratio	dimensionless
η	Viscosity	dyne $cm^{-2}sec$
ϕ	Electric potential	volt
ρQ	Charge density	esu cm^{-3}
τ	MOS transistor transit time	sec
τ	Maxwellian relaxation time	sec
ω	Angular radiation frequency	sec^{-1}

Subscript (all denote components in a mixture)

A	Arbitrary (A = 1, 2, ...)
n or e	Electron
p or h	Hole
i	Ion (in a metal)
a	Atom (in a crystal)

Literature Cited

1. Kittel, C.; Kroemer, H. "Thermal Physics", 2nd Ed., W. H. Freeman: San Francisco, 1980; pp. 355-363.
2. Truesdell, C.; Toupin, R. A. "The Classical Field Theories"; Flugge, S., Ed; ENCYCLOPEDIA OF PHYSICS VOL III/1, Springer-Verlag: Berlin, 1960.
3. D'Avanzo, P. C.; Vanzi, M.; Dutton, R. W. "One-Dimensional Semiconductor Device Analysis (SEDAN)"; Stanford Electronics Laboratory, Technical Report No. G-201-5, Stanford, California, 1979.
4. Selberheer, S.; Schutz, A; Potze, H. IEEE J. Solid State Circuits 1980, 15, 605.
5. Proc. Int'l. Rel. Phys. Symp., 1982, pp. 47-87.
6. Bell, A. E. RCA Rev. 1979, 40, 295.
7. Hall, R. N. Phys. Rev. 1952, 87, 387; Shockley, W.; Read, W. T., Jr. Phys. Rev. 1952, 87, 835.
8. Lifshitz, E. M.; Pitaevski, L. P. "Statistical Physics"; 3rd ed., Part 1, Pergamon: Oxford, 1980 ; p. 167.

9. Landau, L. D.; Lifshitz, E. M. "Theory of Elasticity", 2nd ed., Pergamon: Oxford, 1970; p. 10.
10. Blech, I. A. J. Appl. Phys., 1976, 47, 1203.
11. Blech, I. A.; Herring, C. Appl. Phys. Lett., 1976, 29, 131.
12. Black, J. R. Proc. IEEE, 1969, 57, 1587
13. van Driel, H. V.; Preston, J. S.; Gallant, M. I. Appl. Phys. Lett. 1982, 40, 385.

RECEIVED December 26, 1984

Silicon Oxidation

A Process Step for the Manufacture of Integrated Circuits

Eugene A. Irene

Department of Chemistry, University of North Carolina, Chapel Hill, NC 27514

The manufacture of integrated circuits is essentially microfabrication in two dimensions as it is presently practiced. The microfabrication aspect arises simply from the marketing requirement for a high density of high performance devices on a manufactured chip. The manufacturable device density has risen dramatically in the last ten years with a equally dramatic decrease in the cost per device. An idea of the magnitude of this progress is obtained when we consider the fact that very powerful personal computers are available to consumers at easily affordable costs. Some presently available personal computers approximate the performance of main frames of the 1960's. The packing density of present day devices is primarily determined by the lateral dimensions that can be reliably produced by lithography. Of course some electronic effects arise when the devices get smaller and closer viz. latch-up and hot electron effects, but clever engineering tactics have thus far prevented these effects from retarding further progress. Processing in a direction normal to the substrate surface is relatively easier to control. In this direction one must consider the doping depth and sharpness of impurity profiles in the substrate, and above the substrate the film thicknesses and quality are crucial. The film thicknesses, and in many devices, the depths of dopants in the substrate are usually small relative to the lateral device dimension, hence arises the two dimensional aspect of the manufacturing process. Moreover, present manufacturing practice is limited to one device per area of Si, however, vertical integration schemes are probably not far off and are receiving considerable research effort. The review to follow is aimed at addressing one aspect of the integrated circuit manufacturing process, viz., the preparation of a multipurpose silicon dioxide film on the surface of silicon. The present discussion is focused on silicon oxidation, which is at present the best way to produce a thin film of silicon dioxide, SiO_2. In integrated circuit processing schemes this step is repeated many times. Each repetition addresses a different manufacturing issue in

0097–6156/85/0290–0031$06.00/0

the process. In the discussion that follows, each of these issues will be addressed. Each repetition requires an alteration of the basic oxidation process, but always with the same end, viz. the preparation of a film of SiO_2. We will discuss these processes as well as other techniques (other than thermal oxidation) to produce films of SiO_2. It is crucial to successful manufacturing that the silicon oxidation process be well understood. To this end there is ongoing research in the field and we will discuss this research. Finally, the future needs and requirements for thin films in the manufacture of integrated circuits will be addressed.

Why Silicon Oxidation?

In order to become oriented to the importance of the silicon oxidation process step, it is necesary to consider the fundamental principles of electronic device operation. The great revolution in integrated circuitry took place as it was discovered that devices could be constructed on the surface of a semiconductor. This discovery of how to utilize a semiconductor surface rather than the bulk is, in this author's opinion, the single most important reason for the advancement of electronics technology and the simultaneous emergence of Si as the premier semiconductor. Indeed, Si is mediocre as a semiconductor in terms of bulk semiconducting properties such as carrier mobility. However, for Si, it was found that the surface electronic states resulting from the termination of the crystal lattice could be reduced to tolerable levels by exposure of the surface to oxidation conditions[1]. Fig.1 illustrates that the surface of Si has a large number of unsatisfied bonds, so-called "dangling bonds", as a result of the termination of the lattice. The number of such bonds is of the order of the number of surface atoms which is approximately 10^{15} cm-2. It was predicted that these dangling bonds would produce an equal number of electronic states in the band gap of silicon[2]. This number of states is of the same order or greater than the number of current carriers on the surface of a semiconductor. Therefore, it was correctly reasoned that unless these states were drasticly reduced, the surface electronic properties of the semiconductor could not be used for device construction. The predicted number of states on Si was indeed verified, and more importantly, it was discovered that the number could be reduced several orders of magnitude by simply exposing the surface to the atmosphere and therefore permitting a native surface oxide to form[1]. Of even greater importance was the finding that the purposeful formation of adventitious oxide by the exposure of the Si to an oxidizing ambient at high temperatures would reduce the number of surface states by five orders of magnitude. This process of reducing the number of surface electronic states to tolerable levels is in the electronics parlance termed "passivation" as contrasted with the meaning within the field of electrochemistry. Although there exists considerable research on the subject of electronic passivation, only for Si has the passivation been adequate, and therefore only for Si has the inte grated circuit technology developed to such an advanced state. In fact, even within Si technology other film deposition methods have been attempted such as physical and chemical deposition techniques and these have been shown to be useful for some purposes, but none has yet supplanted thermal oxidation of Si. The approximate comparative results from other techniques as well as for Si oxidation are shown in Table I. The conclusion is that the oxidation process is fundamental to modern integrated circuit technology and we have therefore answered the question "Why" posed above.

Table I.   Comparison of Surface States Resulting from Different SiO_2 Preparation Methods

SiO_2 Preparation Technique	Approximate Number of Surface Electronic States (per area)
Thermal Oxidation	10^{10}
Chemical Vapor Deposition (CVD)	$10^{11}-10^{12}$
Physical Vapor Deposition (PVD)	$10^{11}-10^{13}$
No SiO_2 on Si	10^{15}

For the case of integrated circuit technology there are other very important uses for preparing SiO_2 on Si. It is clear that the electronic passivation is the most important and fundamental use for SiO_2 and that the other uses to be discussed are bonuses, and in many instances could be accomplished equally well and sometimes better with other materials and other processes than SiO_2 and oxidation.

Other Uses For Silicon Dioxide on Silicon

If we consider the physical properties of SiO_2 itself, then several of the other uses become obvious. SiO_2 is a wide band gap insulator (about 9eV). As such, the electrical influence of SiO_2 on the conduction process on the Si surface is nil except for the reduction of the dangling bonds as discussed above. Thus one obtains the electronic passivation without any troublesome interference. The SiO_2 film, being a good insulator, will be able to support a rather large electric field (greater than 10^6 V/cm). Such an electric field applied across the oxide film will alter the Si surface potential and thereby modulate the conduction of carriers in conductive channels created at the Si surface. This effect is the operational principle for the field effect transistor, FET, which when constructed from a metal contact to an oxide film on a semiconductor surface is then called a MOSFET as shown in Fig.2.

Also, when one wishes to prepare a high density of devices on a semiconductor, the isolation of each device from the adjacent device is important. It is interesting to consider that the molar

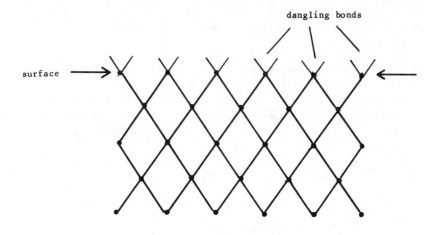

silicon (100) surface

Figure 1.   Side View of (100) Si Single crystal, showing
             dangling bonds at the surface.

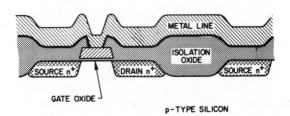

Figure 2.   Metal – Oxide – Silicon field effect transistor,
             MOSFET.

volume of SiO_2 is 120% larger than the molar volume of Si. This
fact can be used to provide one method of isolation. Consider the
adjacent MOSFET devices shown in Fig.2. If one can selectively
oxidize the regions between the devices, the thickness of oxide
formed will rise above the unoxidized regions by more than 100%.
The selective oxidation can be accomplished with the use of an
oxidation mask. Typical of such a masking material is a film of
silicon nitride.

In order to delineate device regions for the purpose of
doping, structural etching of the substrate to fabricate mesa or
recessed structures and to define conducting lines, a masking film
is required. SiO_2 films are often used for all these masking
operations except for oxidation masking. Si_3N_4 is often used for
this purpose. However, Si_3N_4 requires a different method of
preparation called chemical vapor deposition, to be discussed later
and it is known to cause damage to the Si substrate as a result of
a large tensile intrinsic stress(3).

The existence of a film of a hard stable SiO_2 film on Si
serves to protect the Si from damage and impurities.

In summary, over and above the primary use of SiO_2 as an
electrical passivating film, four additional uses of SiO_2 have been
identified. These uses are: electrical operation, isolation,
masking, and protection of the Si.

How is SiO_2 Prepared?

As indicated by the title, the method of thermal oxidation of Si is
presently the most important method to produce SiO_2 films,
especially for critical applications where the number of surface
electronic states and surface charge is important. Before focusing
on the details of that process, it is useful to consider alternate
methods to produce useful films for microelectronics applications.
Other than thermal oxidation of Si, the most important method to
prepare passivating films on semiconductors is chemical vapor
deposition, CVD. CVD utilizes volatile species that either alone or
by reaction produce the desired material in thin film form on a
substrate. For example consider the production of SiO_2 films by
CVD. SiH_4 is often used as the volatile source of Si. The SiH_4 is
reacted with a source of oxygen. Typical choices are CO_2, N_2O, O_2
and H_2O, although the latter two are quite virulent and therfore
usually avoided. The volatile reactants are mixed and possibly
diluted in a furnace above $300°C$ and which also contains the Si
substrates to be coated. With the ability to control the reactant
concentrations, temperature, and total flow rates, uniform thin
films with reasonably good properties can be prepared for all but
the critical interfacial applications where very low surface state
densities are required. Such CVD films are quite useful for masking
operations of all descriptions, for isolation of adjacent devices
and for protecting devices from exposure to contamination. However,
without other treatments, the CVD process cannot produce the low
surface state levels required in active device regions such as the
gate region of a MOSFET. Yet a wide variety of films can be
prepared for an equally wide variety of applications by CVD.
Oftentimes, the basic CVD process is modified to utilize such
energy sources as plasmas, radio frequency induction heating, or

photon sources. In addition to chemical vapor methods are the
physical vapor deposition, PVD, techniques. Evaporation and
sputtering are the basic PVD techniques. Evaporation simply
utilizes a source of desired material plus a source of energy to
vaporize the desired material. Substrates are positioned so as to
receive the recondensed material. Sputtering utilizes accelerated
ions to eject species from a source material, which can then
recondense on a substrate. Both PVD techniques require vacuum
conditions for successful application. Many variances of the basic
PVD techniques are found in actual processes. For evaporation, the
source of energy includes resistance, induction and electron beam
heating of the source, while plasmas and reacive ambients are
commonly employed in sputtering. PVD techniques can be used for
metals, semiconductors, and dielectrics in applications where low
surface state and interface charge densities are not required as
was the case for CVD. With this information about the various other
methods used to prepare thin films for microelectronics, we are
ready to consider the details of the thermal oxidation step and are
able to identify the essential differences among the process steps.

The Thermal Oxidation Process

With the aid of Fig.3 the operational features of the Si oxdiation
process are understood. A source of thermal energy is required, and
this is usually a resistance heated tubular furnace. Since the
process is thermally activated and large batches of substrates are
usually processed at a time, resistance heating is preferred,
because it provides a large and level temperature zone. Cleaned Si
slices are placed on cleaned fused silica carriers and pushed into
the hot zone of the furnace. The furnace is lined with a high
purity fused silica tube. Pure O_2 or H_2O or a mixture are the usual
oxidation ambients. The desired SiO_2 film thickness is grown by
controlling the temperature, time and oxidation ambient. Once
oxidation is complete a short high temperature anneal is required
to reduce interface charge to desired levels. The simplicity of
this process and the emphasis on cleanliness are important
characteristics which are appreciated from a consideration of the
model for Si oxidation. A double wall furnace liner tube is
preferred in order to minimize impurity diffusion through the tube
at high temperatures.

The Silicon Oxidation Model

Firstly, it is important to realize that the Si oxidation model is
a subject of active research at many laboratories around the world.
The information to be discussed about the model is in this authors
opinion, commonly accepted, however there exists differences with
the details, as is commonly the case in any active research area.
Such detail and controversy will be avoided in this review.
There is universal agreement that the rate of thermal
oxidation of silicon decreases as the SiO_2 film grows(4-6). It has
been established that the oxidation reaction that produces SiO_2
occurs at the $Si-SiO_2$ interface(7-9). From all the available
information, the SiO_2 formed in an impurity free environment at

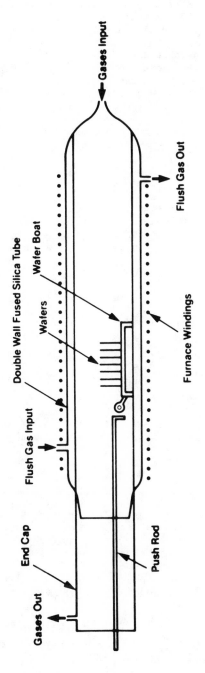

Figure 3. A Basic Silicon Oxidation System.

temperatures below $1200^{\circ}C$ is amorphous, conformal and adherent to
the Si surface regardless of the film thickness. These observables
have led the early workers in this field (see for example refs 4-6)
to consider oxidation by a linear-parabolic process (10). This
linear-parabolic, L-P, model is understood using Fig.4. In this
figure one sees two predominant fluxes. The first, F_1, is the
Fickian diffusion of oxidant across the SiO_2 film to the $Si-SiO_2$
interface, where the reaction with Si produces a flux of growing
SiO_2 , F_2. The transport of oxidant to the outer SiO_2 surface
through the gas phase and the dissolution of oxidant in SiO_2 has
been observed to be rapid under usual oxidation conditions, and
therefore these fluxes can be safely ignored. The fluxes, F_1 and F_2
must be in a steady state, since one flux controls the other as
either the supply or sink. The diffusion flux can be represented by
Fick's first law as:

$$F_1 = D \delta C / \delta L$$

where D is the positionally independent diffusivity of oxidant, and
C and L are the concentration of oxidant and film thickness
respectively. The reaction at the interface can be considered as a
first order chemical reaction in oxidant and is written as:

$$F_2 = kC_2$$

where k is the first order rate constant and C_2 is the
concentration of oxidant at the $Si-SiO_2$ interface. From a steady
state approximation where the fluxes adjust to each other we obtain:

$$F = F_1 = F_2.$$

The rate of film formation, dL/dt, is given as:

$$dL/dt = F/\Omega$$

where Ω is the number of oxidant molecules incorporated into a
unit volume of SiO_2. The integration of the above rate equation
yields a linear-parabolic equation of the form:

$$t = AL + BL^2 + const.$$

There have been several similar methods to evaluate the integration
constant(4,11). Since it is commonly accepted that there exists an
initial regime of oxidation in dry O_2 which does not conform to the
L-P model, this author used the boundary condition that the L-P
regime commences at t_o, L_o. With this condition the final L-P
equation is(11):

$$t-t_o = A(L-L_o) + B(L^2-L_o^2)$$

The constants A and B have dimensions of the reciprocals of the
linear and parabolic rate constants, respectively as:

$$A = 1/k_1 \quad ; \quad B = 1/k_p$$

From this analysis it is clear that in order to predict the specific time to produce a desired SiO_2 film thickness, four parameters need to be known viz. t_o, L_o^2, k_1 and k_p. Such information is necessary in order to characterize a process. The constants can be obtained with reasonable accuracy from dense data sets through the use of curve fitting procedures. Two studies(11,12) show that the dense data sets are conveniently obtained using in-situ ellipsometry which observes the oxide growth in the furnace. The thickness-time data are then fitted to a linearized expression obtained by dividing the resultant L-P equation above by $L-L_o$(11). Values for t_o, L_o are obtained from the data set itself using an iterative procedure. For the first iteration, all the data in an L-t set is fitted to the model and t_o, L_o is assumed to be zero. Values for A and B are obtained as well as a statistical goodness-of-fit parameter. For the next iteration, the smallest L,t point is dropped from the set for fitting purposes, but this data point is taken to be t_o, L_o for the iteration. Again, the quality of the fit is recorded. This procedure is repeated through the data set. The quality of fit parameter is seen to improve as the correct value of t_o, L_o is approached and then remain level beyond the best value. Therefore from this procedure the point in L,t space for the onset of the L-P model is obtained along with the rate constants which best describe the oxidation process. In order to reproducibly prepare SiO_2 of the desired thickness for a given requirement, we must be able to assess the quality of the oxidation model itself. Considerable research has been focused on this problem and this work will be briefly summarized.

How Good Is The Model?

First, we must consider that there is the initial regime which does not conform to L-P kinetics. By the fitting procedures outlined above, this regime was found to extend to 40nm at oxidation temperatures around $1100\,^{\circ}C$ to 5nm at $800\,^{\circ}C$ for oxidation in dry O_2 and to essentially zero for oxidation using H_2O as the oxidant(11,13). These results are in reasonable agreement with other studies that used less dense data sets and graphical fitting procedures(4). At the present time there is no satisfactory model for the initial regime.

For the data thicker than L_o, we can use the linear and parabolic rate constants, k_1 and k_p, to monitor the effects of changes on the specific process in the L-P model viz. interface and/or transport phenomena. The main perturbations to the rate constants arise from the use of different ambients such as H_2O or dry O_2, and different oxidation temperatures.

The effects of H_2O on oxidation kinetics are well documented(5,6,13,14). A systematic study of H_2O effects using dense data sets and the curve fitting procedures outlined above have shown a rather sharp increase in k_p with H_2O concentration and a smoother increase in k_1(13) as shown in Fig.5. Even trace amounts

Linear - Parabolic Model

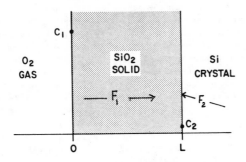

Figure 4.   Oxidant - Silicon Dioxide Film - Silicon System
with Diffusive Flux, F_1, and Reaction Flux, F_2.

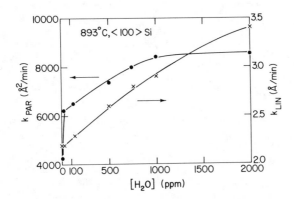

Figure 5.   Linear, k_l, and Parabolic, k_p, Rate Constants
as a Function of the Water in the Oxidation
Ambient.

of H_2O were shown to have profound effects on the kinetics(14). For
example 25 ppm H_2O in O_2 has been found to increase the overall
rate of oxidation by about 20%. The effects have been attributed to
two operative phenomena. One is the effect of the H_2O itself on the
reaction at the Si-SiO$_2$ interface. H_2O is known to be a more
virulent oxidant than O_2, and therefore we anticipate a rate
enhancement due to H_2O, and indeed, an enhancement of k_1 has been
observed. The other effect is related to the fact that H_2O reacts
with SiO$_2$ and breaks up the Si-O-Si network forming hydroxyl
groups. The overall effect is a loosening of the network from which
a diffusivity enhancement should be observed. Indeed, a sharp
increase in k_p is observed with only small amounts of H_2O in the
oxidation ambient as is expected for a structural alteration. Other
impurities such as Na and HCl have also been studied(6,15-17) with
respect to kinetic effects. While the results are not as clear to
relate to the specific rate constants as for H_2O, both of these
impurities increase the overall rate of oxidation. Na is
judiciously avoided during oxidation as it produces mobile positive
charges in the oxide that alter the Si surface potential and
thereby diminish device performance. HCl, on the other hand, is
purposely added to the oxidation ambient to remove any Na that may
have entered the processing environment. Hence, the kinetic effects
for this material need to be known even though complete
understanding is lacking. Available studies yield considerable
information that is useful for processing(5,15-17).

The effect of oxidation temperature on the rate constants in
the L-P regime is shown in Fig.6(18). It is seen that there is
considerable curvature in the Arrhenius plots. This curvature
indicates that neither of the steps yielding the fluxes, F_1 and F_2,
represent kinetically simple rate steps. This finding should not
really be surprising, as the L-P model is quite simplistic.
However, it should be remembered that the L, t data fit the L-P
equation to better than 10% error(11), which is quite respectable
for kinetics models, and the model is physically satisfying .The
conclusion concerning the L-P model is that the model very
adequately represents the oxidation kinetics for the L-P regime.
The rate constants are used industrially to define processes and as
starting parameters for process design.

It should be clear that there are real conceptual problems
with the L-P model notwithstanding its practical utility. The
existence of a regime of oxide thicknesses that does not conform to
the model plus the evidence for more complex steps has opened new
research avenues and the quest for an even more satisfactory model.

A Revised Oxidation Model

In addition to the evidence above that demonstrates the
inadequacy of the L-P model, there exists further information which
with the above evidence requires some revision for the L-P model.
Before the evidence is presented, it is useful to consider the
mechanical circumstances that exist at the Si-SiO$_2$ interface during
oxidation. The molar volume of SiO$_2$ is 120% larger than that of Si.
Thus we can envision that when Si converts to SiO$_2$, volume must be
found to permit the reaction to proceed. For normal chemical
reactions taking place with finely dispersed materials in the gas

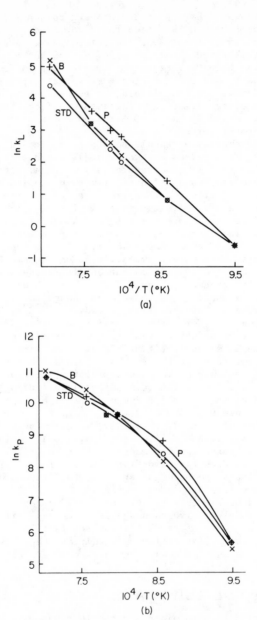

Figure 6.   Arrhenius plots of the linear (a) and parabolic (b) rate
constants.   (Reproduced with permission from Ref. 18.
Copyright 1978 The Electrochemical Society.)

or solution phases this situation usually presents no difficulty
However, for the case of a rigid thin film that strongly adheres to
the rigid substrate, mechanical consequences arise. For the
particular case of Si with a growing high molar volume SiO_2 film,
we would anticipate the accumulation of an intrinsic compressive
film stress (tensile in the Si). When this stress is large enough,
it could actually oppose the foward chemical reaction. However,
prior to the cessation of the reaction, the stress could cause
defect formation in the Si and/or delamination of the oxide. The
fascinating observation is that for oxidation above about $1000^{\circ}C$,
we observe no intrinsic stress, no damage to Si and no
delamination. Below about $900^{\circ}C$, an intrinsic stress is observed
that increases with decreasing temperature, an increase in oxide
density, oxide fixed interface charge, and increased surface states
all of which increase with decreasing oxidation temperature(19-22).
Of course as discussed above, there is also a change in the
oxidation kinetics below $1000^{\circ}C$. All of these phenomena can be
qualitatively explained using the concept of viscous flow in the
oxide.
      The observation of viscous flow in SiO_2 films was first
reported by EerNisse(23,24). Essentially, a compressive intrinsic
stress was found to exist in SiO_2 films grown below $1000^{\circ}C$ and this
stress was relieved at higher temperatures. The densification of
SiO_2 films was reported(20,21) and a unified model that explains
both the occurrence of stress in SiO_2 and the higher density was
published(25). This model utilized the concept of viscous
relaxation in a Maxwell solid. The main idea is depicted in Fig.7
where the molar volume change is seen to cause the stress and
density increase which are both relieved via viscous flow at
sufficiently high temperatures. These ideas were recently
incorporated into a revised oxidation model(26). Part of this
revision modifies the oxidation expression to include the stress
driven viscous relaxation. From a consideration of SiO_2 as a simple
Maxwell solid the expression for F_1 becomes:

$$F_1 = k' C_2 C_{Si} \sigma / \eta$$

where σ and η are the oxide stress and viscosity. A new rate
constant is defined which does not include the orientation
dependence that is now made explicit through C_{Si}.
      This new way of explaining the low temperature kinetics may
have great technological relevance, since processing demands
require lower processing temperatures and thinner SiO_2 and
therefore it is crucial that we   understand the mechanical
effects of the oxidation process.

Alternative Oxidation Technologies

Because of manufacturing demands for lower process temperatures,
interest in two alternative oxidation techniques has been
rekindled. One technique is high pressure oxidation. From the L-P
model we can see that an increased oxidant pressure will accelerate
transport across the growing film and thereby increase the rate.
This technique was identified in the 1960's (27) but was ahead of

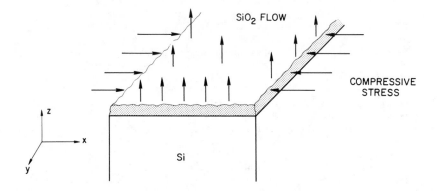

Figure 7.   A viscous flow model for silicon dioxide film formation.
(Reproduced with permission from Ref. 25.   Copyright 1982
The Electrochemical Society.)

its time then. In the middle 1970's it was reintroduced(28) and
properties of the resulting oxides studied(21). Commercial
equipment is presently in use at many major semiconductor producing
companies, and the benefits of this technique have been discussed.
The basic kinetics and film formation mechanism seems to be similar
to normal 1 atm SiO_2 growth(21,28,29) and the film properties also
appear similar to 1 atm oxides grown at the same low temperature as
the high pressure oxides. However, substantial data proving these
assertions is still lacking. The other novel technique is plasma
oxidation. An oxygen plasma is brought in contact with the silicon
surface. The plasma contains many small and highly energetic
species that apparently greatly enhance the rate of oxidation. This
technique was also introduced for Si oxidation in the 1960's(30),
and also ahead of its time then. Newer studies are not in agreement
as to the mechanism(31,32). High oxidation rates are achievable
especially when anodization occurs by placing a potential on the
Si. It appears that some form of high temperature treatment is
required to bring the interfacial properties in line with the thermal
oxide. Research is active in both these areas. Of less importance
presently, but perhaps of future significance, is the use of high
intensity light from lasers to enhance the oxidation rate(33-35).
Even less is known about this process than the other two. Much
research is presently taking place on these techniques and the
future manufacturing requirements will undoubtedly stimulate the
search for yet more and better processing methods.

Conclusions

In the preceeding discussion, the thermal oxidation of Si was shown
to be a valid and integral process step in the manufacture of
integrated circuits. Only this process produces the highest quality
SiO_2 for critical device operation. Other methods that also produce
SiO_2 in thin film form, and that are also valid process steps were
contrasted with thermal oxidation and the utility of the steps was
disccussed. The oxidation model was shown which yields considerable
predictability to process engineers. Problems with the model were
pointed out, viz. low temperature oxidation and thin film growth.
These growth regimes are becoming technologically very important.
In order to provide a more suitable model research is underway at
several laboratories as well as the exploration of several novel
film growth techniques. Integrated circuit manufacturing demands
will continue to drive research at improving the thermal oxidation
process step.

Literature Cited

1. Atalla, M.M.; Tannenbaum, E.; Scheibner, E.J.
    Bell System Tech. J., 1959, 38, 749.
2. Shockley, W. Phys. Rev., 56, 1939, 317.
3. Irene, E.A. J. Electron. Mater., 1976, 5, 287.
4. Deal, B.E.; Grove, A.S. J. Appl. Phys., 1965, 36,
    3370
5. Pliskin, W.A. IBM J. Res. and Develop., 1966, 10,
    198.

6. Revesz, A.G.; Evans, R.J. J. Phys. Chem. Solids, 1969, 30, 551.
7. Ligenza, J.R.; Spitzer, W.G. J. Phys. Chem. Solids, 1960, 44, 131.
8. Pliskin, W.A.; Gnall, R.P. J. Electrochem. Soc., 1964, 111, 872.
9. Rosenscher, E.; Straboni, A.; Rigo, S; Amsel, G. Appl. Phys. Lett, 1979, 34, 254.
10. Evans, U.R. "The Corrosion and Oxidation of Metals"; Arnold, London, 1960, Chap. XX.
11. Irene, E.A.; van der Meulen, Y.J. J. Electrochem. Soc., 1976, 123, 1380.
12. Hopper, M.A.; Clarke, R.A.; Young, L. J. Electrochem. Soc., 1975, 122, 1216.
13. Irene, E.A., Ghez, R., J. Electrochem. Soc ., 1977, 124, 1757.
14. Irene, E.A. J. Electrochem. Soc., 1974, 121, 1613.
15. Hess, D.W.; Deal, B.E. J. Electrochem. Soc., 1977, 124, 736.
16. van der Meulen, Y.J.; Cahill, J.G. J. Electronic Mater., 1974, 3, 371.
17. Deal, B.E. J. Electrochem. Soc., 1978, 125, 576.
18. Irene, E.A.; Dong, D.W. J. Electrochem. Soc., 1978, 125, 1146.
19. Deal, B.E.; Sklar, M.; Grove, A.S.; Snow, E.H. J. Electrochem. Soc., 1967, 114, 266.
20. Taft, E.A. J. Electrochem. Soc. 1978, 125 968.
21. Irene, E.A., Dong, D.W., Zeto, R.J. J. Electrochem. Soc., 1980, 127, 396.
22. Nicollian, E.H. J. Vac. Sci. Technol., 1977, 14, 1112.
23. Eer Nisse, E.P. Appl. Phys. Lett., 1979, 35, 8.
24. Eer Nisse, E.P. Appl. Phys. Lett., 1977, 30, 290.
25. Irene, E.A.; Tierney, E.; Angilello, J. J. Electrochem. Soc., 1982, 129, 2594.
26. Irene, E.A. J. Appl. Phys., 1983, 54, 5416.
27. Ligenza, J.R. J. Electrochem. Soc., 1962, 109, 73.
28. Zeto, R.J.; Thornton, C.G.; Hryckowian, E.; Bosco, C.D. J. Electrochem. Soc., 1975, 122, 1409.
29. Lie, L.N.; Razouk, R.R.; Deal, B.F. J. Electrochem. Soc., 1982, 129, 2828.
30. Ligenza, J.R. J. Appl. Phys., 1965, 36, 2703.
31. Ho, V.Q.; Sugano, T. Jap. J. Phys., 1980, 19, 103.
32. Ray, A.K.; Reisman, A. J. Electrochem. Soc., 1981, 128, 2466.
33. Schafer, S.A.; Lyon, S.A. J. Vac. Sci. Technol., 1981, 19, 494.
34. Boyd, I.W. Appl. Phys. Lett., 1983, 42, 728.
35. Young, E.M.; Tiller, W.A. Appl. Phys. Lett., 1983, 42, 63.

RECEIVED March 12, 1985

Convective Diffusion in Zone Refining of Low Prandtl Number Liquid Metals and Semiconductors

William N. Gill[1], Nicholas D. Kazarinoff[2], and John D. Verhoeven[3]

[1] Chemical Engineering Department, State University of New York at Buffalo, Amherst, NY 14268
[2] Mathematics Department, State University of New York at Buffalo, Amherst, NY 14268
[3] Materials Science and Engineering Department, Iowa State University, Ames, IA 50011

Several elementary aspects of mass diffusion, heat
transfer and fluid flow are considered in the context
of the separation and control of mixtures of liquid
metals and semiconductors by crystallization and
float-zone refining. First, the effect of convection
on mass transfer in several configurations is considered from the viewpoint of film theory. Then a
nonlinear, simplified, model of a low Prandtl number
floating zone in microgravity is discussed. It is
shown that the nonlinear inertia terms of the momentum
equations play an important role in determining
surface deflection in thermocapillary flow, and that
the deflection is small in the case considered, but it
is intimately related to the pressure distribution
which may exist in the zone. However, thermocapillary
flows may be vigorous and can affect temperature and
solute distributions profoundly in zone refining, and
thus they affect the quality of the crystals
produced.

The basic idea of zone refining is that a liquid region or zone is
created by melting a small fraction of the material in a relatively
long solid charge, ingot or feed stock. By moving this liquid zone
through the charge in one direction the solid phase can be purified
as the forward surface is melting and the rear one solidifies.
This is referred to as zone refining. If the liquid zone is passed
through the charge in both the forward and reverse directions, a
uniform distribution of impurities may be obtained. This is zone
leveling.

Pfann (1) first described the essential features of zone
refining and pointed out its potential as a separation technique.
In the early 1950's it was used to provide high purity silicon and
germanium for semiconductor applications. Since then it has been

0097–6156/85/0290–0047$06.50/0

used in a variety of applications. The real power of zone refining is that one can pass a molten zone through a solid phase numerous times without difficulty. Up to a limit each pass increases the purity of the solid phase by decreasing the concentration of the solute. It is by means of multipass operation that great purity can be achieved in the solid phase.

Zone refining is a dynamic nonequilibrium separation process. However, to understand its essential features it is important to understand the equilibrium concepts on which it is based. Of central importance is the notion of the equilibrium distribution coefficient, $k_0 = C_S/C_L$, where C_S and C_L are the concentrations of solute at equilibrium in the solid and liquid phases respectively. Since the value of k_0 may be dramatically different from unity, it is clear that at equilibrium a solute may distribute itself between solid and liquid phases with a great preference for one or the other at a given temperature. On this basis the relative amount of solute in each phase can be controlled, and a separation can be carried out.

Phase diagrams, which describe the equilibrium relations that exist between phases in mixtures, are often very complex as is the case with the nickel-aluminum diagram for example. However, if one restricts attention only to a relatively narrow range of concentration and, in particular, to dilute solutions, then one can simplify this description of these relationships markedly by using straight line approximations as shown in Figures 1(a, b).

Figures 1(a, b) represent phase relationships for cases in which solid solutions exist, a common occurrence in metallic and semiconductor systems. In such cases solidification does not cause complete separation and the degree of separation depends not only on the equilibrium relationships represented by Figures 1(a, b) but also on the convective-diffusive characteristics of the system. If the addition of solute lowers the melting point as in Figure 1a, then $k_0 < 1$; on the other hand if the melting point is raised by adding solute, $k_0 > 1$ as in Figure 1b.

A solute distribution exists in the melt because the solidification is carried out at a finite rate. For example, if $k_0 < 1$, then solute is rejected and accumulates at the surface which is solidifying, and this creates solute gradients in the melt which tend to be relaxed by molecular diffusion and any convection which may exist. The <u>interfacial</u> distribution coefficient, k, refers to the solid to liquid solute concentration ratio at the interface. It is k which is used in transport calculations when one is trying to understand the dynamic behavior of zone refining systems. It usually is found that equilibrium exists locally at the solid-liquid interface, in which case $k = k_0$.

The concentration, C_L, of solute in the liquid in the neighborhood of the solid-liquid interface is strongly influenced by k which thereby affects the concentration in the solid phase after solidification in at least two ways. First, the smaller k is the faster solute will accumulate on the liquid side of the interface. Second,

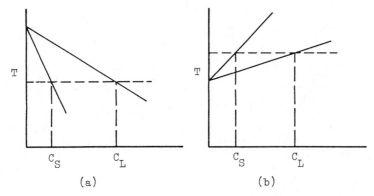

Figure 1. Possible solid-liquid phase diagrams.

it is the value of $C_L(0)$ which determines C_S at the interface for
a given value of k. With no convection in the melt, it will be
shown that the maximum interfacial concentration is approached as
the process progresses and is, $C_L(0) = C_L(\infty)/k$, where $C_L(\infty)$ is the
concentration of solute in the solution outside the diffusion
boundary layer. $C_L(0)$ climbs from $C_L(\infty)$ to the $C_L(\infty)/k$ asymptote.
    The material to be purified often is very reactive, and it is
difficult to find container materials which it will not attack. In
such cases floating-zone melting is attractive and is used for
example to grow oxygen free silicon crystals. In this case the
molten zone is held in place by its own surface tension which works
against the action of gravity. A molten zone can be established
between two rods, one a polycrystalline charge or feed rod and the
other a crystalline rod of purified material which is either the
product of the separation or may serve as the feed for another pass.
The molten zone may be created in several ways including radio-
frequency induction heating as is used for silicon or electron beam
heating which is favored for refractory materials.
    The gravitational field limits the length of the molten zone
that can be sustained by surface tension forces, and this also
limits the crystal diameter. The possibility of reducing these
constraints by decreasing significantly the intensity of the
gravitational field has generated considerable interest in
performing floating zone experiments in an orbit around the Earth.
    Heating from an external source creates rather large tempera-
ture gradients along the surface of the floating zone which give
rise to strong thermo-capillary convective flow which mixes the
melt. Furthermore this flow may couple with natural convection
flows driven by the gravitational field. These flows give rise to
time-dependent behavior which is of considerable theoretical and
practical interest because it creates growth striations in semiconductor crystals which affect their performance, and because its
origin, and the parameters which characterize it, are not well
understood.
    A detailed review of zone melting and its applications has been
given recently by Shaw (2). In the present paper we shall confine
our attention primarily to the convective-diffusive characteristics
of such systems, and we shall strive primarily to obtain a sound
qualitative understanding of their behavior.
    The flow phenomena involved in zone refining will be discussed
briefly. In particular we shall consider surface tension driven
flow in a cavity containing a low Prandtl number, Pr, fluid (a low
Pr number is typical of liquid metals and semiconductors). It will
be shown that simplified models of such flow, which simulate the
melt configuration in zone refining, predict multiple steady-state
solutions to the Navier-Stokes equations exist over a certain range
of the characteristic parameter.

The Equations of Motion, Energy and Diffusion for Molten Zones

In actual zone refining operations one may encounter three-
dimensional simultaneous heat and mass transfer with moving interfaces, and the system also may be time dependent. The mode of
heating affects system behavior significantly, as does radiation,

when materials with high melting points are involved. All of these
factors complicate the analysis of the diffusional aspects of this
separation process. Therefore we shall treat some rather simplified
versions of real systems in an effort to gain some insight into
several of the important phenomena which occur.

First, we shall assume the system is two-dimensional and that
the physical properties, except surface tension and density (in the
body force term of the momentum equations) are constant. With these
assumptions the continuity equation, momentum equations, energy
equations and the diffusion equation are given by

$$\frac{\partial u}{\partial x} + \frac{\partial v}{\partial y} = 0 \tag{1}$$

$$\frac{\partial u}{\partial t} + u\frac{\partial u}{\partial x} + v\frac{\partial u}{\partial y} = -\frac{1}{\rho_\infty}\frac{\partial P}{\partial x} + \upsilon\left[\frac{\partial^2 u}{\partial x^2} + \frac{\partial^2 u}{\partial y^2}\right] + \frac{\rho}{\rho_\infty}g_x \tag{2}$$

$$\frac{\partial v}{\partial t} + u\frac{\partial v}{\partial x} + v\frac{\partial v}{\partial y} = -\frac{1}{\rho_\infty}\frac{\partial P}{\partial y} + \upsilon\left[\frac{\partial^2 v}{\partial x^2} + \frac{\partial^2 v}{\partial y^2}\right] + \frac{\rho}{\rho_\infty}g_y \tag{3}$$

$$\frac{\partial T}{\partial t} + u\frac{\partial T}{\partial x} + v\frac{\partial T}{\partial y} = \alpha\left[\frac{\partial^2 T}{\partial x^2} + \frac{\partial^2 T}{\partial y^2}\right] \tag{4}$$

$$\frac{\partial C}{\partial t} + u\frac{\partial C}{\partial x} + v\frac{\partial C}{\partial y} = D\left[\frac{\partial^2 C}{\partial x^2} + \frac{\partial^2 C}{\partial y^2}\right] \tag{5}$$

The boundary conditions for Equations 1-5 depend on the mode of
heating and the way the separation is conducted. To gain some
insight into how the separation occurs at the solid-liquid interface
we first consider Equation 5. We shall assume that the melt is
quiescent so that the process is governed entirely by diffusion and
u and v are zero. If there is no convection in the melt and
directional solidification is occurring only in the x-direction,
then Equation 5 becomes

$$\frac{\partial C}{\partial t} = D_L \frac{\partial^2 C}{\partial x^2} \tag{6}$$

Let us now assume, as is customary, that the charge is being moved
through the heater at a constant velocity U, and that this causes
the solid-liquid interface to move at the velocity U. Then it is
convenient to use a coordinate X which moves with the interface such
that

$$X = x - Ut \tag{7}$$

and Equation 6 becomes

$$\frac{\partial C}{\partial t} - U\frac{\partial C}{\partial X} = D_L\frac{\partial^2 C}{\partial X^2} \qquad (8)$$

At time $t = 0$ the solute concentration in the charge was C_0.
Since diffusion in the liquid phase is slow, we can assume the diffusion layer near the interface is thin compared to the length of
the melt; and therefore $C(t,\infty) = C_0$. Now we must consider the
boundary condition at the interface. To do this we equate the
fluxes in the solid and liquid phases at $X = 0$ and remember that
with respect to X there is an apparent convective velocity equal to
$-U$. Furthermore, $D_L \gg D_S$ so that solid phase diffusion can be
neglected. Therefore

$$-\frac{\partial C}{\partial X}(t,0) = \frac{(1-k)\ U}{D_L}C(t,0) \qquad (9)$$

where $k = \dfrac{C_S(t,0)}{C_L(t,0)}$. If we assume $k = k_0$ and use the idealized phase
diagram with straight lines shown in Figure 1, k is constant and
this makes it much easier to solve the problem posed in Equation 8
and 9 together with the other initial and boundary conditions.

Order of Magnitude Considerations

There are various ways of using Equations 8 and 9 to obtain information about the solidification process. The simplest one is to do
an order of magnitude analyses, OMA, of these equations. This
yields immediately that on a relative basis the first, second and
third terms are of order $1/t$, U/δ and D_L/δ^2, where δ is the approximate thickness of the diffusion boundary layer. Equating the first
and last terms gives

$$\delta \sim \sqrt{D_L t} \qquad (10)$$

and the second and last terms give

$$\delta \sim D_L/U \qquad (11)$$

A similar OMA of Equation 9 yields

$$\frac{C(0)}{C_0} \simeq \frac{1}{1 - (1-k)\dfrac{U\delta}{D_L}} \qquad (12)$$

If one combines Equations 10, 11 and 12, one gets

$$\frac{C(0)}{C_o} = \frac{1}{1 - (1-k)\ tU^2/D_L} \qquad t \ll \frac{D_L}{U^2} \qquad (13)$$

and

$$\frac{C(0)}{C_o} = \frac{1}{k} \qquad t \gg \frac{D_L}{U^2} \qquad (14)$$

Equations 13 and 14 provide a qualitative picture of the relationship between the concentration at the solid-liquid interface and
that in the bulk of the melt.

   We see from Equations 10-14 that two time regimes exist in the
problem. The first period is a transient period which exists at
$t \ll D_L/U^2$. The second period is a steady state one which occurs
when $t \gg D_L/U^2$. Since D_L is a physical property which cannot
be manipulated, we see that the greater the velocity of solid-liquid
interface the more quickly a steady state is reached in the system.
This conclusion is of practical importance because one often wishes
to distribute the solute throughout the solid with a constant
concentration.

Exact One-Dimensional Solutions of Diffusion Equation

In the steady state $\partial C/\partial t = 0$, and Equations 8 and 9 can be solved
easily to give

$$C_L/C_o = 1 + \frac{1-k}{k}\ \exp\left(-\frac{UX}{D_L}\right) \qquad (15)$$

which gives the steady state solution for the interfacial concentration as

$$\frac{C_L(0)}{C_o} = \frac{1}{k} \qquad (16)$$

Equation 16 shows clearly that the <u>solid</u> phase steady concentration
is C_0, the initial melt concentration. It is interesting to note
that Equations 14 and 16 are identical.

   The solution of Equation 8, if one includes the unsteady state
effect is more complex, and was apparently first shown by Smith et
al. (3) to be,

$$\frac{C_L}{C_o} = 1 - \frac{1}{2}\left\{ \mathrm{erfc}\ \frac{X+Ut}{2\ \sqrt{D_L t}} + \frac{hD_L}{hD_L-U}\ e^{\frac{-UX}{D_L}}\ \mathrm{erfc}\ \frac{X-Ut}{2\ \sqrt{D_L t}} \right.$$

$$+ \frac{U-2hD_L}{2(U-hD_L)} \ e^{-hX + ht(hD_L-U)} \ \text{erfc} \ \frac{X+(U-2hD_L)t}{2\sqrt{D_L t}}\} \qquad (17)$$

where $h = (1-k)U/D_L$.

Equation 17, evaluated at $X = 0$, together with the definition of k, and setting $t = x/U$, enables one to calculate the solid phase distribution as

$$\frac{C_S}{C_o} = \frac{1}{2}\{1 + \text{erf}\sqrt{\frac{Ux}{4D_L}} + (2k-1) \ \exp\left[-k(1-k)\frac{Ux}{D}\right] \ \text{erfc}\left[(2k-1)\sqrt{\frac{Ux}{4D_L}}\right]\} \qquad (18)$$

where x is the distance from the point at which the first solid was frozen.

Effect of Convection on Segregation

The preceding discussion assumes that no convection exists in the melt, and this is rarely, if ever, the case. Next we shall consider two approaches which account for convection in the melt, a transport mechanism which is especially important in mass transfer because D_L is small and even weak convection markedly alters solute concentration profiles and may cause macrosegregation. First we shall discuss film theory which is a very simple approach that gives qualitative information and often provides considerable physical insight into the mechanisms involved. Second, we shall discuss a simplified model of zone refining.

In film theory one assumes that the concentration changes near the interface occur in a very thin region of thickness δ. In this region the differential equation which describes the concentration is given by the steady state $\left(\frac{\partial C}{\partial t} = 0\right)$ form of Equation 8. The magnitude of δ is determined by the convection which exists in the system. One estimates δ by solving simplified convection problems such as natural or forced convection along a flat surface, or two or three-dimensional stagnation flow, etc. From these simple cases one calculates the mass transfer coefficient, k_m, and δ is defined as $\delta = D_L/k_m$. In essence, one neglects all convection when calculating C in the stagnant film, and includes all convective effects in its thickness, δ. The stronger the convection the smaller is δ.

The steady-state solution satisfying Equations 8 and 9 and $C = C_o$ at $X = \delta$ is

$$\frac{C}{C_o} = \frac{k + (1-k)\,e^{-\frac{U}{D_L}X}}{k + (1-k)\,e^{-\frac{U}{D_L}\delta}} \qquad (19)$$

Therefore, one obtains the well known Burton–Prim–Slichter (4) equation

$$\frac{C_s}{C_o} = \frac{k}{k + (1-k)e^{-\frac{U}{D_L}\delta}} \qquad (20)$$

and the problem is reduced to choosing an expression for δ which represents a realistic estimate of the convective patterns which exist in the melt. Let us use a few relevant examples to illustrate how film theory may be applied to zone refining.

By definition, the mass transfer coefficient is given by

$$j = -D_L \frac{\partial C(0,y)}{\partial x} = k_m (C(0) - C_o)$$

where y is the distance parallel to the surface and x is normal to it. Therefore the definition of δ is,

$$\delta = \frac{D_L}{k_m} = \frac{C(0) - C_o}{-\frac{\partial C}{\partial x}(0,y)} = \frac{y}{Sh_y}$$

where Sh_y is the Sherwood number defined by $\frac{k_m y}{D_L}$ and $\frac{\partial C(0,y)}{\partial y}$ is calculated from a problem similar to the one to which we are applying film theory, but one that can be analyzed more easily. For example, to calculate $\frac{\partial C}{\partial x}(0,y)$, one would not include the moving boundary because that is included in the film model, the steady state form of Equation 8. One might also use a constant concentration or constant gradient boundary condition at the interface rather than Equation 9 which also is included in the film model.

Suppose natural convection is the dominant convection mode in the melt. If a natural convection boundary layer created by concentration gradients exists, then its behavior depends on whether the interface is vertical or horizontal to the earth. If it is vertical, one can show that,

$$Sh_y = 0.54 \, Ra_y^{1/4} = 0.54 \left(\frac{\beta g \Delta C y^3}{\nu D_L} \right)^{1/4} \qquad (21)$$

may be a reasonable approximation, Kays and Crawford ($\underline{5}$), and

$$\delta \cong \frac{1}{0.54} \left(\frac{\nu D_L y}{\beta g \Delta C} \right)^{1/4} \qquad (22)$$

If it is horizontal, Stewartson ($\underline{6}$) and Gill et al. ($\underline{7}$) showed that

$$Sh_y \simeq 0.75 \ Ra_y^{1/5} \tag{23}$$

may apply if the interface faces upward and the density of the fluid adjacent to it increases with distance from the interface; or if it faces downward and the density relationship is reversed. In this case

$$\delta \simeq \frac{4}{3}\left(\frac{\nu \ D_L \ y}{\beta g \ \Delta c}\right)^{1/5} \tag{24}$$

The important point made by Equations 20, 22 and 24 is that δ is a function of y, the distance parallel to the interface, and this leads to segregation in the y direction (a nonuniformity of solute concentration) in the solid phase as given in Equation 20.

It is most desirable for δ to be constant, and there are ways to make this happen. If the solid-liquid interface is circular, which is most often the case because the charge is a rod, then δ can be controlled by rotating the rod so that the interface behaves as a rotating disk with angular velocity ω. This configuration is widely used in analytical and electrochemistry because δ is essentially constant if this mechanism controls. Levich (8) has shown that for a rotating disk

$$\delta = 1.61 \left(\frac{D_L}{\nu}\right)^{1/3} \sqrt{\nu/\omega} \tag{25}$$

Equation 25 shows that δ is constant and its magnitude can be controlled by changing ω.

Natural convection to blunt bodies such as cylinders (2-dimensional) and spheres (3-dimensional) has been studied by Acrivos (9) and from his analysis one can show that these configurations are characterized by constant boundary-layer thicknesses. For 2-dimensional bodies,

$$\delta = \frac{1}{0.54} \left(\frac{R \ \nu \ D_L}{\beta g \ \Delta C}\right)^{1/4} \tag{26}$$

and for 3-dimensional ones,

$$\delta = \frac{1}{0.54(2)^{1/4}} \left(\frac{R \ \nu \ D_L}{\beta g \ \Delta C}\right)^{1/4} \tag{27}$$

where R is the radius of curvature of the surface. Note that Equation 26 is identical to 22 except that the length scale is R which is a constant, and thus δ is a constant. This implies that the interfacial concentration of solute is uniform across the stagnation region. These results may apply to an interface which faces downward (in the direction of the gravity vector) into a fluid whose density increases with distance from the interface or one

which faces upward into a fluid of decreasing density with distance
from the interface.

When strong temperature gradients exist, natural convection may
be primarily induced thermally or both heat and mass transfer may
play comparable roles. In these cases the situation is more
complex, because the number of parameters increases. In liquid
metals and semiconductors the Schmidt number, ν/D_L, is several
orders of magnitude greater than the Prandtl number, ν/α, and this
enables one to solve for the concentration profile in a rather
general way without great difficulty as will be discussed next.

The number of alternative configurations of the melt and modes
of heating it that may exist in zone refining is very large. Therefore it seems desirable to have a method which enables one to estimate macrosegregation under a wide variety of flow conditions. Such
an approach is not available now, but some progress toward it can be
made by noting that liquid phase diffusion is characterized by a
large Schmidt number, which implies that diffusion boundary layers
are thin compared to momentum boundary layers. It seems that
Lighthill (10) was the first to suggest that one can restrict
attention to the velocity field near the interface under these conditions, and by doing this one can derive the expression

$$\delta = 0.894 \left(9 D_L \mu\right)^{1/3} \tau_w^{-1/2} \left[\int_o^y \tau_w^{1/2} \, dy \right]^{1/3} \tag{28}$$

Equation 28 is a general result for high Sc, two-dimensional flows
in which the diffusion boundary layer thickness is zero at $y = 0$.
It includes Equations 22 and 24 as special cases, but it does not
apply to systems in which $\delta \neq 0$ at $y = 0$, such as stagnation
regions, and also it does not include Equation 25. Equation 28 can
be applied to flows which are driven by temperature differences
regardless of the magnitude of the Prandtl number.

To use Equation 28 for thermally driven free convection
boundary layers one simply calculates $\tau_w = -\mu \frac{\partial u}{\partial x}(0,y)$ from known
solutions. Then this result is inserted in Equation 28 and δ in
Equation 19 or 20.

Unfortunately, because of the variety of factors, such as
shape, mode of heating and orientation, that is possible in melts,
and the complexity of the flow patterns which may exist, it is
extremely difficult to offer general rules, a priori, on how to
estimate δ. One needs to examine each particular case carefully to
obtain even a qualitative understanding of the macrosegregation that
may occur in the crystals being produced. However, one procedure
which seems to yield generally beneficial results is crystal rotation as predicted by Equation 25.

Simplified Model of Surface Tension Driven Flow in a Two-dimensional Molten Zone Supported on the Bottom

The preceding discussion showed that steady-state natural convection
often leads to undesirable macrosegregation. It also has been shown
by Gill (11) that natural convection flows may become unstable, and

Carruthers (12) and Milson and Pamplin (13) have discussed the
implications of the resulting oscillations on crystal growth. In
this section we shall examine exact solutions of the Navier-Stokes
equations for a two-dimensional simplified model of a molten zone
which is in the form of a cavity or slot of liquid of depth d,
supported on the bottom, but with a free surface on top. The zone
is heated over the length, $-\ell < x < \ell$, by a flux, q, and cooled on
its ends at $x = \pm L$, where $L > \ell$. We shall study the core region
inside $-\ell < x < \ell$ for which a similarity solution exists. Thus we
are neglecting end-effects.

One can show that Equation 4 is satisfied by a temperature
distribution in the form

$$\theta = \frac{T - T_{cold}}{T_{hot} - T_{cold}} = g_1(\eta) + g_2(\eta)\left(\frac{x}{\ell}\right)^2 \qquad (29)$$

where g_1 and g_2 are functions which are determined from

$$g_1'' + 2A^3 Ma\ f\ g_1' = -2A^2 g_2 \qquad (29a)$$

and

$$g_2'' + 2A^3 Ma\left[f\ g_2' - 2\ f'\ g_2\right] = 0 \qquad (29b)$$

with the initial conditions (See Appendix I)

$$g_1'(0) = g_2'(0) = 0 \qquad (29c)$$

$$g_1(0) = -g_2(0) = 1 \qquad (29d)$$

and $\eta = y/d$. Here $f(\eta)$ and $f'(\eta)$ are functions related to u and v
which will be determined later; Ma is the Marangoni number
$\frac{\Delta T\ \ell |d\alpha/dT|}{\mu\ \sigma}$; and A is the aspect ratio $\frac{d}{\ell}$. In the limit Pr $\to$ 0, the
solution given by solving Equations 29 implies a constant heat flux
along the bottom of the cavity, $\eta = 1$, and a zero heat flux from the
free surface, $\eta = 0$, into the vacuum surrounding the liquid zone.
For non-zero Pr the flux at $\eta = 1$ varies with x. Obviously Equation
29 implies that the liquid surface temperature varies as

$$T(x,0) = T_h - (T_h - T_c)\left(\frac{x}{\ell}\right)^2 = T_h - \Delta T x^2_1 \qquad (29e)$$

where T_h is the temperature at $x = 0$ and T_c is that at $x = \pm\ell$.
Gill et al. (14) have shown by numerical computation that Equation
29e is a good approximation to a constant heat flux for fluids with
finite values of Pr which are typical of liquid metals. The following discussion applies to all Pr fluids, but low Pr is the category
that includes essentially all fluids of interest in semiconductor
technology as well as all liquid metals. On the other hand, the
most complete data on thermocapillary flows in molten zones has been
reported by Preisser, Scharmann and Schwabe (15) and Schwabe and

Scharmann (16) for NaNO$_3$ which has a Prandtl number of 9 and for heating from the side which is not consistent with the present theory.

Levich (8) has discussed capillary motion in two-dimensional creeping flows in which the surface was flat. Yih (17) pointed out inconsistencies in Levich's analysis which were associated with the assumptions of a linear distribution of surface tension with distance along the interface, and with the deflection of the surface which inevitably occurs when capillary flow exists. He noted that under certain circumstances steady flows may not exist. Ostrach (18, 19) has discussed scaling problems in capillary flows. Recently, Sen and Davis (20) studied capillary flow in bounded cavities in which d/ℓ is small, end effects are present, and the flow is very slow and the cavity is heated from the side. Cowley and Davis (21) studied the high Marangoni number "Thermocapillary analogue" of a buoyancy driven convection problem solved by Roberts (22). Later, we shall make some comparisons between our results for the deflection of the surface and those of Sen and Davis (20).

The boundary conditions for Equations 1–3 which will be satisfied at x = 0 and the solid bottom are straightforward, but those for the free surface are rather complex and call for some discussion. At x = 0 we have a stagnation condition and at y = d there is no slip and no penetration of fluid through the solid bottom. Therefore

$$u(x, d) = v(x, d) = 0, \ (0 \leqslant x \leqslant \ell) \text{ and } u(0, y) = 0, \ (0 \leqslant y \leqslant d) \quad (30)$$

The free surface is not flat in general and the boundary conditions on this surface require careful consideration. The kinematic condition at y = -h(x), where h(x) is the deflection from the flat surface at y = 0, is

$$v(x, -h) = -\frac{dh}{dx} u(x, -h) \quad (31a)$$

Equation 31a is the steady state form of the kinematic condition which also is used to describe wave motion as discussed on page 595 of Levich's book, Physicochemical Hydrodynamics. One also must equate the normal and tangential components of the forces in each phase at the free surface. Since we consider a gas-liquid interface, we neglect gas phase resistance due to its viscosity and include only the pressure it imposes on the interface on the gas side. Therefore, at y = -h(x) the tangential component of the stress tensor for the liquid phase is equal to the tangential force created by the change in surface tension with temperature in the x direction. Thus the tangential force balance at the interface becomes

$$\frac{-\mu}{1 + h_x^2} \left[(1-h_x^2)\left(\frac{\partial u}{\partial y} + \frac{\partial v}{\partial x}\right) - 2h_x\left(\frac{\partial v}{\partial y} - \frac{\partial u}{\partial x}\right) \right] = \frac{\sigma_x}{\left(1 + h_x^2\right)^{3/2}} \quad (31b)$$

and the normal force balance becomes

$$-(p-p_g) + \frac{2\mu}{(1 + h_x^2)} \left[\frac{\partial v}{\partial y} + h_x\frac{\partial u}{\partial y} + h_x(-h_x\frac{\partial u}{\partial x} + \frac{\partial v}{\partial x})\right] = \frac{\sigma\, h_{xx}}{(1 + h_x^2)^{3/2}} \quad (31c)$$

where σ is the surface tension and the x-subscripts on h and σ denote differentiation with respect to x, and p_g is the pressure in the gas phase over the liquid which at zero g is taken to be zero.

To solve Equations 1-3 together with 30 and 31a-c we take note of the important role played by the surface tension number, $\varepsilon = \frac{\mu\nu}{d\sigma_0}$, where σ_0 is the surface tension at temperature T_0. In the liquid metal and semiconductor systems we are considering ε is very small (i.e. $\varepsilon \simeq 10^{-7}/d$ for liquid tin); therefore a first order perturbation solution in ε should be adequate. Consequently we introduce a stream function, ψ and a similarity coordinate, η, so that Equations 1-3, 30 and 31a-c are satisfied to $O(\varepsilon)$, and the computations are reduced to solving a two-point-boundary-value problem for a single nonlinear ordinary differential equation in the dimensionless stream function f. The procedure for finding the base flow solution, which is equivalent to finding f and η for $\varepsilon = 0$, is similar to that employed by Carter and Gill (23) to treat the entirely different physical problem of flow with heat transfer, blowing or suction, and natural convection in flows between parallel-plates or in tubes. As usual we define

$$u = \frac{\partial\Psi}{\partial y} \qquad -v = \frac{\partial\Psi}{\partial x} \qquad (32)$$

We choose $\eta = y/d$ which suggests

$$\Psi = U_R(x)\, d\, \int_o^{\eta} \frac{u(x,y)}{U_R(x)}\, d\eta$$

If $u(x,y)$, scaled by $U_R(x)$, is a function of η only, then

$$\Psi = U_R(x)\, d\, f(\eta) \qquad (33)$$

Equations 2, 32 and 33 lead to

$$f''' + \frac{d^2}{\nu}\frac{dU_R}{dx}\left(ff'' - f'^2\right) = \frac{d^2}{\mu U_R}\frac{\partial p}{\partial x} = \beta \qquad (34)$$

which satisfies Equations 1-3 exactly. Clearly the coefficient $\frac{d^2}{\nu}$

$\frac{dU_R}{dx}$ is constant, and Equation 31b is satisfied, if

$$U_R = \frac{2\left|\frac{d\sigma}{dT}\right| \Delta T d}{\mu \ell} \left(\frac{x}{\ell}\right) \quad (34a)$$

so that

$$u = 2\left(\nu/d\right) A^2 Re\, x_1\, f'(\eta), \quad v = -2\left(\nu/d\right) A^3\, f(\eta). \quad 35(a,b)$$

where $x_1 = x/\ell$ and $A = d/\ell$. To obtain Equation 34a we use Equation 29 and $\sigma = \sigma_0 + \frac{d\sigma}{dT}(T-T_0)$ which yields $\sigma/\sigma_0 = 1 + ARe\,\varepsilon\, x_1^2$. Equation 34 then becomes

$$f''' + 2\left(\frac{d}{\ell}\right)^3 \frac{Ma}{Pr}(ff'' - f'^2) = \beta \quad (36)$$

and the boundary conditions become

$$f(0) = f(1) = f'(1) = f''(0) + 1 = 0 \quad (37)$$

where $\frac{Ma}{Pr} = Re = \frac{\Delta T\, \ell\, |d\sigma/dT|}{\mu\alpha}\left(\frac{\alpha}{\nu}\right)$ and the last boundary condition given by Equation 37 is obtained by substituting Equations 32a,b into Equation 31b which is satisfied to $O(\varepsilon)$. Since Equation 36 is of third order and 37 includes four boundary conditions, the constant β can be determined as a function of $Q = 2\left(\frac{d}{\ell}\right)^3 \frac{Ma}{Pr}$, and the solution for f can be related to u and v via Equations 32 and 33. An important feature of Equation 36 is that it retains the essential nonlinear features of the Navier-Stokes equations. It will be seen later that the nonlinear terms are of great importance even at relatively small values of Q, and it is they which determine if multiple solutions to Equation 36 do exist.

Brady and Acrivos (24) studied an entirely different physical problem with different boundary conditions from those given in Equation 37 and cautioned that the numerical solutions they obtained for a system with a capped end differed from the similarity solutions. Their problem differs fundamentally from the present one in several ways. First, the flow in our problem is symmetric about x = 0, because surface heating is symmetric about x = 0 over the region $-\ell < x < \ell$, and this causes u = 0 at x = 0. Hence, our solution is exact near x = 0. Their problem is antisymmetric about x = 0, and they apparently have a "collision region" near x = 0 which affects the structure of the entire flow. Second, Brady and Acrivos specify the shape of their cavity, h(x), as part of the problem statement, and their choice of shapes may be inconsistent with the pressure distribution in their similarity solutions which

they use as the criterion for agreement with their numerical calculations. We solve for the shape of the zone and require it to be
consistent with the pressure distribution to $O(\varepsilon)$ in Equation 31c.
Third, the shape of the zone affects the boundary conditions on v
through the kinematic condition, and therefore our boundary conditions are different from theirs. Furthermore, Gill et al. (14)
found good agreement between similar solutions for thermocapillary
flow in an infinite disk and numerical solutions to a finite disk
with end effects. In the thermocapillary problem (14) the boundary
conditions were identical to those in Equation 37.

   We studied the two-point-boundary-value problem (TPBVP)
Equations 36 and 37 (corresponding to a slot with one free surface)
by suspending the trivial differential equation

$$\beta' = 0 \qquad\qquad (38)$$

to 36 and using the TPBVP solver BOUNDS, P. Deuflhard (25), to solve
the resultant well-posed problem, Equations 36, 37 and 38. BOUNDS
is an accurate, multiple-shooting code. We began with the solution

$$f = \eta(1-\eta)^2/4, \quad \beta = 3/2 \qquad\qquad (39)$$

which corresponds to $Q = 0$. We found a family of solutions for
$Q > 0$ evolving from Equation 39; we first slowly increased Q and
then increased Q in larger increments up to approximately $Q = 8000$.
For continuation we used linear interpolation of the solutions
obtained for Q_1 and Q_2 ($Q_1 < Q_2$) to provide the guess for the solution
at $Q = Q_3 > Q_2$. Above $Q = 8000$, β began to change more rapidly, and
we replaced (38) by

$$Q' = 0 \qquad\qquad (40)$$

in order to prescribe β and compute Q and f. We continued by linear
interpolation (as with Q) until $\beta = -0.75$ where $Q \approx 20,000$. A
summary of the numerical results is provided in Figure 2.

   For $Q \lesssim 6900$ the solutions had 2 cells; see Figure 2. Above Q
≈ 6900, a third cell appeared and increased in strength as β was
decreased; see Figure 2. A local maximum in Q was reached at $Q =$
8614.48, $\beta = -0.0675$; a local minimum in Q was reached at $Q =$
8128.32, $\beta = -0.21$. Thus for $8128.32 < Q < 8614.48$ we found three distinct solutions of the TPBVP, Equations 36 and 37. Figure 3 compares the three solutions of $Q = 8600$ with that for $Q = 500$. It is
seen that the differences between the velocity distributions on the
middle and lower parts of the branch are so small as to be indis-
tinguishable on the figure.

   The rapid change in β with Q for $0 < Q < 10^3$ shows clearly that
nonlinear effects are important in Equation 36 even when the
Marangoni number is relatively small.

   Now let us calculate the deflection of the surface and
determine how it depends on the parameters of the system. To do
this, substitute Equations 35a,b into 31c, together with

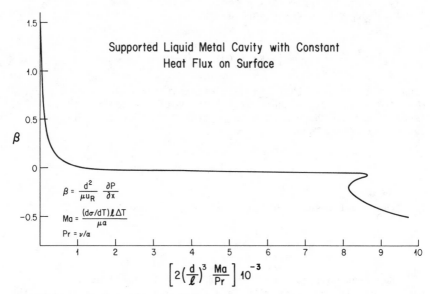

Figure 2.   Results of numerical evaluation of TPBVP, Eqns. 36 and 37.

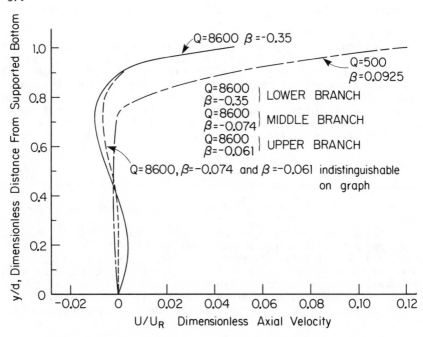

Figure 3.   Numerical predictions of axial velocity versus height above the support surface.

$h = h_0 + h_1\epsilon$, and by equating coefficients of ϵ one obtains $h_0 = 0$ and for $\bar{H}_1 = h_1/d$ the equation

$$\frac{d^2 H_1}{dx_1^2} = \left[4ARe\ f'(0) + \frac{p_0 d^2}{\mu\nu} + \frac{\beta}{A}\ Re\ x_1^2\right] \tag{41}$$

with the conditions that the deflection at the ends is zero and that the liquid just fills the cavity. To avoid the complications associated with the end-effects we take $\ell = L$ for the purpose of calculating H_1. Thus

$$H_1(-\ell) = H_1(\ell) = \int_{-1}^{1} H_1\ d\ x_1 = 0. \tag{42}$$

Then the final result for h is

$$h/d = \frac{\beta}{60}\frac{Re}{A}\ \epsilon\left(1 - 5x_1^2\right)\left(1 - x_1^2\right) \tag{43}$$

Equation 43 shows that the deflection is small because ϵ is small, that it increases with Re/A and that its sign depends on the sign of β. However, it must exist in order to make the pressure distribution given by $\dfrac{d^2}{\mu U_R}\dfrac{\partial p}{\partial x} = \beta$ consistent with the boundary condition given by Equation 31c.

Sen and Davis (20) have performed an interesting analysis of capillary flow in a slot heated from the side (a half zone) by using the matching procedure suggested by Cormack, Leal and Imburger (26) for natural convection in a slot. In this way they take end-effects into account. However, they assume that the nonlinear terms in the momentum equation can be neglected and perform an analysis for small values of d/ℓ, which they use as their perturbation parameter. Their result, to the first order, is

$$\frac{h}{d} = \frac{Re}{8A}\ \epsilon\left[x_1 - 3x_1^2 + 2x_1^3\right] \tag{44}$$

where x_1 is the dimensionless distance from the hot wall in their case.

One cannot compare Equations 43 and 44 quantitatively because they apply to different configurations and therefore the distribution of h with x is different. However, it is interesting to compare these results qualitatively. Let us look at the main point of similarity first. In the limit Re $\to$ 0, $\beta = 3/2$, and for this value of β, Equation 43 predicts a maximum deflection of 0.025 Reϵ/A while Equation 44 predicts 0.012 Reϵ/A. Obviously, the parameter,

$$\text{Re}\,\varepsilon/A = \frac{\left|\frac{d\sigma}{dT}\right|\,\Delta T}{\sigma_o\,(d/\ell)^2}$$ on which the deflection depends is the same in both

cases. We note also that this parameter is inversely proportional to $(d/\ell)^2$. Thus the smaller (d/ℓ) is, the larger the fractional deflection, h/d. On the other hand in the linear flow case Equation 43 predicts a larger deflection than Equation 44; this may be due to the neglect of end-effects in our case or the fact that heating from the side is studied by Sen and Davis and we have considered surface heating. The most important difference between Equations 43 and 44 is the appearance of β, which depends on the parameter $2(d/\ell)^3$ Re and therefore reflects the nonlinear terms in the Navier-Stokes equations. For a fixed (d/ℓ) one sees from Figure 2 that h/d first decreases with Re because β decreases rapidly, and then h/d increases with Re because β remains relatively constant up to Re ~8000.

Figure 2 shows that values of β change rapidly from positive to negative as $2\left(\frac{d}{\ell}\right)^3$ Re increases from zero to a little over 1100. Beyond a value of $2\left(\frac{d}{\ell}\right)^3$ Re ~ 6900 a third cell appears, and at about 8100 multiple solutions for β appear after it has changed very slowly from about 1100 to 8000. The rapid changes in β at small values of $2\left(\frac{d}{\ell}\right)^3$ Re suggest that nonlinear effects play a decisive role in the behavior of the deflection, h/d, in Marangoni flow.

We speculate that the existence of multiple solutions to the Navier-Stokes equations for surface tension driven flow in a cavity is evidence of instability in this system, and this suggests the need for a stability analysis of the velocity profiles. It should be noted that the simplified model of the molten zone which we are using does not include the blocking effect of the solid-liquid interface at $x = \pm L$.

Calculations by Gill et al. (14), in which a similar simplified model was used for a disk shaped molten zone, agreed well with more detailed finite difference calculations, except in the immediate neighborhood of the solid liquid interface. Both types of calculations for the molten disk suggested that only one steady-state similarity solution exists and that β changes rapidly with $2\left(\frac{d}{\ell}\right)^3$ Re so that nonlinear effects also are important in that configuration.

The disk problem was reduced to equations similar to Equations 36 and 37 except, for the disk, the term ff'' is replaced by $2ff''$. Therefore, the term $v\frac{\partial u}{\partial y}$ is weighted more heavily compared $u\frac{\partial u}{\partial x}$ in the disk than in the cavity and this eliminates multiple solutions. This seems to mean that a relative increase in the convection of x-direction momentum normal to the free surface compared to its convection parallel to the free surface favors stability in the zone. However, a stability analysis is required to confirm this hypothesis.

Smith and Davis (27) have carried out an interesting stability analysis on liquid layers in a slot heated from the side. They neglected end effects and the nonlinear inertia terms in their base-flow velocity distribution and identified three types of instabilities: stationary roll cells, hydrothermal waves and surface waves. In systems with return flows, as is the case in floating or supported zones, stationary roll cells were shown not to exist in slots heated from the side. Surface waves were found to be the prevailing cause of instability at very low Pr. However, it seems tenuous to try to extrapolate their results to the return flows with significant inertial effects and surface heating which we studied because the nonlinear inertia terms have a profound influence on the velocity profile, the surface deflection and the pressure distribution; therefore they would influence strongly all modes of instability.

It is clear that vigorous thermocapillary flows with surface velocities on the order of 10 cm/sec may exist in low Prandtl number liquid metal or semiconductor zones. These flows no doubt have a profound effect on the temperature and solution distribution in these zones, and thereby they affect crystal quality. This conclusion is consistent with that drawn by Wilcox and coworkers, (28) and (29), who first called attention to the potential importance of surface tension driven flows in float zone solidification. Schwabe and Scharmann (16) have recently summarized experimental investigations on a wide variety of high Pr number fluids. Surface velocities have been directly measured in transparent fluids with Pr numbers varying from 0.3 for molten KCl to over 1000 for silicon oils and molten glass. The study on molten glass, for example, recently reported by McNeil et al. (30) with both top and bottom heating of a liquid bridge reported flows of roughly equal velocity in the range of 0.1 to 1 cm/s. This study and the several others listed by Schwabe and Scharmann (16) confirm the predominance of surface-tension flow over natural convection even at 1 g, and it seems reasonable that this same conclusion will apply to liquid metals and semiconductors at Pr numbers which fall into the .01 to .03 range.

Appendix

A more general approach to solving the energy equation than that given in (29-29a) can be carried out quite easily to yield various physically meaningful exact solutions of Equation 4 which depend on the mode of heating used. To accomplish this let:

$$T = T_R + \left[g_1(\eta) + g_2(\eta) \left(\tfrac{x}{\ell} \right)^2 \right] \Delta T \tag{A1}$$

where T_R and ΔT are reference temperatures and temperature differences which will be determined later and which depend on the way the zone is heated. Equation A1 obviously satisfies the symmetry condition at x=0. Now let the heat balances at $\eta=0,1$ be

$$-k \frac{\partial T}{\partial y} (x,o) = h_1 \left[Te_1 - T (x,o) \right] \tag{A2}$$

and

$$-k \frac{\partial T(x,d)}{\partial y} = h_2 \left[T(x,d) - T_{e_2} \right] \tag{A3}$$

where Equations A2 and A3 may represent conductive, convective or linearized conditions.

The environmental temperatures that the surfaces see, T_{e1} and T_{e2}, will not be constant. For example, in actual zone refining operations a radiant ring heater which encircles only part of the liquid zone may be used. Therefore, realistic boundary conditions must take this into account. Consequently, we expand $T_{e_i} - T_R$ in a Taylor series in x, reconize that the distribution must be symmetric about x=o so the first derivative vanishes, and truncate the series after the third term. The result is, $\theta_i = \dfrac{T_{ei} - T_R}{\Delta T} = \theta_{i_1} - \theta_{i_2} (\dfrac{x}{\ell})^2$. The constants θ_{ik} reflect the variation in θ_i with x. If $\theta_{i_2} > \theta_{i_1}$, the liquid surface near x=o will be heated and that near the solid-liquid interfaces will be cooled. If $\theta_{i_1} >> 0_{i_2}$, then θ_i is essentially a constant. If one substitutes Equation A1 into A2 and A3 one gets

$$-g_1' (0) = Nu_1 [\theta_{11} - g_1(0)] \tag{A4}$$

$$g_2' (0) = Nu_1 [g_2 (0) + \theta_{12}] \tag{A5}$$

$$g_1' (1) = Nu_2 [\theta_{21} - g_1(1)]$$

$$g_2' (1) = -Nu_2 [g_2(1) + \theta_{22}]$$

where $Nu_1 = \dfrac{h_1 d}{k}$, $Nu_2 = \dfrac{h_2 d}{k}$.

Obviously, the Nu_1 and θ_{1k} are specified as part of the problem statement.

Equations 29a, b are linear. Therefore, each of them may be solved for any two sets of boundary conditions and one then can construct by superposition of these new solutions for arbitrary boundary conditions without additional computation. For example, let

$$g_1 = a_1 h_1 (\eta) + b_1 h_2 (\eta) \tag{A8}$$

$$g_2 = a_2 j_1 (\eta) + b_2 j_2 (\eta) \tag{A9}$$

where the $h_i (\eta)$ and $j_i (\eta)$ satisfy Equations 29a, b for arbitrarily chosen boundary conditions. The a_i and b_i now easily can be chosen to satisfy A4 to A7. Furthermore, by comparing Equations 29 and A1 we see that

$$\Delta T = - (\dfrac{T_H - T_c}{g_2(o)})$$

and

$$T_R = T_H + \dfrac{g_1(o)}{g_2(o)} (T_H - T_c)$$

where $g_1 (o)$ and $g_2(o)$ are found in general from A8 and A9.

This solution also offers an approximation for the shape of the solid-liquid interface as:

$$(\frac{x}{\ell})^2 = \frac{g_1(o) + \frac{T_H - T_m}{T_H - T_c} g_2(o) - g_1(\eta)}{g_2(\eta)}$$

If $T_c = T_m$ = melting point, this reduces to:

$$(\frac{x}{\ell})^2 = \frac{g_1(o) + g_2(o) - g_1(\eta)}{g_2(\eta)}$$

These estimates of solid-liquid interface shape and position are based on u and v which satisfy all boundary conditions except the no slip condition at the solid-liquid (melting-freezing) interfaces. Therefore, to determine their accuracy, they must be compared with other solutions based on velocity fields which do satisfy these conditions.

Acknowledgments

The work described in this paper was done under a grant from the National Aeronautics and Space administration.

Literature Cited

1. Pfann, W. G. Trans. Met. Soc. AIME 1952, 194, 747.
2. Shaw, J. S. "Zone Melting and Applied Techniques" in "Crystal Growth"; B. R. Parryslin, Ed., Pergamon Press, New York, 1975.
3. Smith, V. G.; Tiller, W. A.; Rutter, J. W. Can. J. Phys. 1955, 33, 723.
4. Burton, J. A.; Prim, R. C.; Slichter, W. C. J. Chem. Phys. 1953, 21, 1987.
5. Kays, W. M.; Crawford, M. E. "Convective Heat and Mass Transfer"; 2nd Ed., McGraw Hill, New York, 1980.
6. Stewartson, K.; ZAMP 1958, 9a, 276.
7. Gill, W. N.; Zeh, D. W.; del Casal, E. ZAMP 1965, 16, 539.
8. Levich, V.; "Physicochemical Hydrodynamics"; Prentice Hall, Inc.: Englewood Cliffs, N.J., 1962.
9. Acrivos, A.; A I Ch E Journal 1960, 6, 584.
10. Lighthill, M. J. Proc. Rog. Soc. 1950.
11. Gill, A. E. J. Fluid Mech. 1974, 64, 577.
12. Carruthers, J. R. J. Crystal Growth 1976, 32, 13.
13. Milson, J. A.; Pamplin, B. R. Prog. Crystal Growth Charact. 1981, 4, 195.
14. Gill, W. N.; Kazarinoff, N. D.; Hsu, C. C.; Noack, M. A.; Verhoeven, J. D. "Thermocapillary Driven Convection in Supported and Floating Zone Crystallization"; COSPAR, Advances in Space Research: Pergamon Press, 1984.
15. Preisser, F.; Schwabe, D.; Scharmann, A. J. Fluid Mech. 1983, 126, 545.

16. Schwabe, D.; Scharmann, A. "Microgravity Experiments on the Transition from Laminar to Oscillatory Thermocapillary Convection in Floating Zones"; COSPAR, Advances in Space Research: Pergamon Press, to be published in 1984.
17. Yih, C. S. Phys. Fluids 1968, 11, 477.
18. Ostrach, S. "Physicochemical Hydrodynamics"; V. C. Levich Festschrift (ed. D. B. Spalding) 1977, 2, 571.
19. Ostrach, S. Ann. Rev. Fluid Mech. 1982, 14, 313.
20. Sen, A. K.; Davis, S. H. J. Fluid Mech. 1982, 121, 163.
21. Cowley, S. J.; Davis, S. J. Fluid Mech. 1983, 135, 175.
22. Roberts, G. O. Geophys. Astrophys. Fluid Dyn. 1977, 8, 197 and 1979, 12, 235.
23. Carter, L; Gill, W. N. AIChE Journal 1964, 10, 330.
24. Brady, J; Acrivos, A. J. Fluid Mech. 1981, 112, 127 and 1982, 115, 427.
25. Deuflhard, P. "Recent Advances in Multiple Shooting Techniques" in Computational Techniques for Ordinary Differential Equations, Caldwell and Sayer, Eds, Academic Press, New York, 1980.
26. Cormack, D. E.; Leal, L. G.; Imberger, J. J. Fluid Mech. 1974, 64, 577.
27. Smith, M. K.; Davis, S. H. J. Fluid Mech. 1983, 132, 119 and 145.
28. Chang, C. E.; Wilcox, W. R. Int. J. Heat Mass Trans. 1976, 19, 335.
29. Clark, P. A.; Wilcox, W. R. J. Crys. Gr. 1980, 50, 461.
30. McNeil, T.; Cole, R.; Subramanian, S. "Surface Tension Driven Flow in Glass Melts and Model Fluids" in Materials Processing in the Reduced Gravity of Space, Guy Rindone, Ed, Elsevier, 1982, p. 289-299.

RECEIVED December 26, 1984

Research Opportunities in Resist Technology

David S. Soong

Department of Chemical Engineering, University of California, Berkeley, CA 94720

Selected research areas in resist technology requiring
knowledge of polymer physics, solution thermodynamics,
heat and mass transfer considerations, optics and
rheology, are discussed. First, the success of
lithographic processes depends on the reproducible
generation of desired polymer resist film thickness
and uniformity. A large number of process variables
affect the outcome of spin coating. Mathematical
modeling is instrumental in elucidating the dominant
mechanism governing film formation. The non-Newtonian
character of the resist solution must be taken into
account, as well as changes in resist viscosity and
solvent diffusivity during spinning, since the polymer
concentration increases due to solvent evaporation.
Next, high resolution lithography entails the
accurate writing of the desired pattern in the radiation-sensitive material. However, edge definition is
often compromised by electron back-scattering for the
case of e-beam resists and interference of reflected
with incident light to produce standing waves for
photo-resists. Experimental and modeling efforts in
the literature are reviewed.
In-situ monitoring of resist development is also
an active area of research. This is achieved with a
number of techniques, including the use of a Psi-meter
or an ellipsometer in our laboratory. The ability to
record optical properties as functions of time during
film dissolution allows determination of the effects
of process variables (such as developer strength and
resist molecular weight) on dissolution characteristics. Two-dimensional simulation based on knowledge
acquired via these in-situ experiments will then give
some insight to the origin of pattern distortion
induced by polymer-solvent interactions.

Engineering considerations relevant to prebaking
and postbaking are discussed, and potential problems
for modeling are identified. Finally, resist performance under plasma conditions is studied via analysis
of the molecular weight distribution in successive
layers of the film by GPC. Average molecular weights
increase with depth, signifying that the most extensive degradation occurs at the plasma-resist interface. However, the superficial etch rates in different plasma compositions do not correlate with the
molecular weight profiles, suggesting additional
surface erosion mechanisms other than random chain
scission.

Research opportunities in resist technology continue to grow
in many directions, as the performance demands on the resists and
associated processes increase with our desire to miniaturize microelectronics devices. It is a difficult task, if not an impossible
one, to fully review the fundamental engineering considerations of
all the current and future areas of research in this rapidly
evolving field. Hence, the objectives here will be rather limited,
in view of the great diversity of potential research emphases. The
topics selected for discussion reflect my own interests, which are
much influenced by personal experience and past association with
scientific collaborators.

Resist (photoactive or radiation sensitive) materials are
divided into organic and inorganic categories. This discussion
will focus on organic resists that are generally polymeric in
nature. Potential inorganic resists have been reviewed by Tai et.
al. (1). The basic working principle of resist technology hinges
on radiation-induced structural changes, to an extent sufficient to
alter material dissolution characteristics or etch resistance in
subsequent solution or plasma development. A great deal of work
has been done to understand the chemistry of resists. The goal is
to design systems with improved sensitivity, contrast, resolution,
and etch resistance. The current status and challenges of
chemistry-related problems have been reviewed by Willson (2).

Research areas where a chemical engineer can expect to make
the most contributions are primarily in resist processing. Here,
the major engineering considerations often entail basic conservation laws for mass, momentum and energy, polymer physics, thermodynamics, and transport properties in polymeric matrices. An
excellent review has been written by Thompson and Bowden (3) to
address the issues encountered in individual unit operations. A
few of the many topics discussed in their review will be selected
here for a more in-depth examination. Hardware parameters such as
lens imperfections, source stability, mask quality and dimensions,
contamination and mechanical stability do not fall in the scope of
the present discussion. This paper will concentrate on events and
research challenges require a firm understanding of material

behavior. Process optimization resulting from such research will involve changes in operational procedure or process conditions.

This paper, in brief, presents some research opportunities associated with five resist processing steps: spin-coating, exposure, development, baking and etching. The approaches suggested here are geared toward acquiring a fundamental understanding of the interplay of the process parameters with material properties. Such theoretical understanding may eliminate empirical trial-and-error process optimization.

Spin-Coating

Spin coating of resist films on semiconductor substrates is an important step in the fabrication of integrated circuits. Film formation is accomplished by dispensing a fixed amount of polymer solution onto a wafer. The wafer is then rotationally accelerated up to a pre-set speed. The resist solution flows radially outward due to centrifugal force, reducing the fluid layer thickness. Evaporation of solvent from the resist solution continuously changes the fluid composition, and thereby the rheological properties of the fluid. Clearly, momentum transport and diffusion of volatiles in polymers are both important ingredients to a successful description of this process.

Understanding the film formation process is critical to determining the outcome of the entire lithographic process. The need has been intensified in recent years as device features shrink to submicron sizes. Resist film uniformity must be held within a small tolerance to minimize exposure artifacts. Currently, achievement of this requirement demands a significant amount of time and effort to characterize a new resist material so that a viable spin-coating process can be established. Appreciable reduction of this effort could be attained if a quantitative model were available to describe the fluid behavior during spin coating.

A number of theoretical as well as experimental studies of the spin-coating process has been reported in the literature (4-14). Empirical correlations with different process parameters have been generated based on a large number of experiments (7,10,12-14). It has been established that the initial volume of the fluid dispensed onto the rotating disk and the rate of fluid delivery (i.e., dispense speed during a brief period of slow rotation preceding spinning to spread out the fluid) have a negligible effect on the final film thickness. On the other hand, the final film thickness is a strong function of the resist viscosity and final steady spin speed. An increase in angular velocity decreases the film thickness. An inverse power-law relationship generally holds for the thickness dependence on the final spin speed. At a given speed, the final film thickness exhibits a different power-law relationship with the polymer concentration in the resist solution. Increasing the polymer concentration leads to thicker films, presumably because of the increased fluid viscosity. For a given process, the film thickness decreases rapidly at first, but then slows down considerably at longer times (4).

Previous attempts to model the spin coating process have relied on several major simplifying assumptions. Most treatments

have assumed either Newtonian or power-law viscosity behavior with
time-independent rheological properties during the process (4-7,
8-9,11). The neglect of solvent evaporation in the spin-coating
process is a significant limitation of these efforts. During
spinning, solvent evaporation drastically changes the rheological
properties of the fluid, which in turn greatly affect the rate at
which the film thickness decreases with time due to radial flow.
Only in the work of Meyerhofer (11) has an effort been made to
account for the effects of solvent evaporation on resist viscosity. To facilitate the numerical computations, it was assumed
that the process is composed of two limiting steps. At short
times, the film thickness variation due to solvent loss was
ignored. Beyond a certain critical point, it was assumed that film
thickness reduction was due only to evaporation, and not to radial
flow. The transition from one loss mechanism to the other was
approximated by the point at which the rates of the two loss processes become equal.

It is clear that a detailed numerical simulation of the spin
coating process is needed to elucidate the dependence of film
thickness and thickness profile on all controllable process
variables, such as polymer concentration of the resist solution,
polymer molecular weight, polymer-solvent compatibility or solvent
power, initial volume of resist dispensed on the wafer, spinner
acceleration characteristics, duration of acceleration, final
steady-state spinner speed, ambient temperature and solvent saturation level, solvent volatility, diffusivity of solvent in the
film, and, in case a mixed solvent pair is used, the relative composition. Changes in a particular variable often induce changes in
other characteristics of the process. Furthermore, the interrela-
tionships are so convoluted as to prevent clear-cut trends to be
predicted with straightforward arguments. The recognition of this
process complexity and the need for a quantitative model motivated
us to develop a detailed mathematical description of the spin-
coating process (15). The model has been evaluated by comparing
predicted film thicknesses and thickness profiles with those
observed experimentally for a variety of spin-coating conditions.

Important steps in the development of the model are briefly
highlighted below. First, a cylindrical coordinate system
(Figure 1) is chosen to describe the process. The system is
characterized by disk radius, R, and angular velocity, $l(t)$. Film
thickness, w, is a function of radial distance, r. The value of w
changes with time during spinning. In our analysis, fluid flow is
assumed to be rotationally symmetrical, obviating derivatives with
respect to r in conservation equations. Convection is restricted
to the radial direction. This lubrication approximation greatly
simplifies the equations of continuity and motion. The assumption
is supported by the observation that w is not a strong function of
r. Surface tension forces are neglected, as the fluid initially
floods the wafer, and the radius of surface curvature is small only
near the edges. The validity of the model in the immediate
vicinity of a thin ring surrounding the edge of the wafer may not
hold due to high surface tension contributions. For multiple
coatings over pre-existing topography, surface tension forces are
significant near corners of protruded or receded steps.

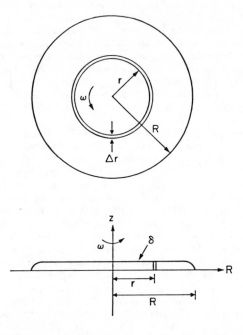

Figure 1.   Overhead and side views of the coordinate system used to describe resist spin coating.   The wafer revolves along the z-axis at an angular speed ω.   The disk has a radius R.   Film thickness, δ, is a function of time and radial position, r.

The continuity equation, the equation of motion, and component
mass conservation equation can be easily written with the above
assumptions. Proper initial and boundary conditions can be imposed. These equations are general and apply to all resist systems. Differences exist in the constitutive equations describing
material and transport properties.

Viscosity of the resist solution as a function of concentration is affected by both free-volume change and entanglement
formation as the solvent evaporates (16). The former can be
accommodated in a manner similar to the development of the Fujita-
Doolittle equation (17, 18), whereas proper consideration of the
latter must invoke scaling concepts (19) to account for chain-chain
interactions. For the present effort, we assumed the viscosity
function to be a product of the above two components:

$$\eta_{po} = k \; c^m \exp \left(\frac{B}{f} \right) \tag{1}$$

where η_{po} is the polymer contribution to the system viscosity in
the limit of low shear rates, c is the polymer concentration, B is
a constant of the order unity, f is the fractional free volume (20)
at the given temperature and concentration (a quantity calculated
by assuming linear additivity of component contributions), and k
and m are material constants. Exponent m undergoes a sudden
transition from about $1 \sim 1.5$ to approximately 4 at a critical
concentration for incipient entanglement formation, depending on
the polymer molecular weight and solvent strength.

The above viscosity relation does not yet take into account
the non-Newtonian feature of the resist solution. When the rotational speed is high, the fluid is sheared to a large extent,
inducing appreciable shear thinning of the material. Hence,
approximation of the fluid behavior by a Newtonian constitutive
equation leads to inaccurate predictions. The Newtonian model not
only predicts too thick a film, but also gives too weak a film
thickness dependence on spinner speed, as will be shown below.

To accommodate non-Newtonian effects, a realistic constitutive
equation is used for the shear-rate dependent viscosity (21-23).
The overall system viscosity is divided into contributions from the
polymer and remaining solvent:

$$\eta = \eta_p + \eta_s = \frac{\eta_{po}}{1 + b \, \dot{\gamma}^a} + \eta_s \tag{2}$$

where η is the total system viscosity, η_p and η_s are the polymer
and solvent contributions, b and a are parameters characterizing
the onset of non-Newtonian behavior and the slope of the power-law
relationship for the shear-rate, $\dot{\gamma}$ dependence. The quantity
η_{po} is the polymer contribution to the viscosity at zero shear and
is described by Equation 1.

The dependence of the solvent diffusivity on the solution
concentration embodies the same free-volume concept as before, with
a temperature dependence built in automatically (24,25). This

diffusivity relation endows strong dependence of diffusion rate on the remaining local solvent concentration. As solvent evaporates, the diffusivity of solvent in the relatively depleted top layer may drop by orders of magnitude. A diffusion barrier is thus set up to protect the bulk fluid from substantial further solvent loss, enabling the still liquid-like resist to flow outwards.

The resist system chosen for experimental study was a standard electron beam resist, poly(methyl methacrylate), (PMMA). Resist solutions of 4%, 6% and 9% PMMA in chlorobenzene were purchased from KTI Chemicals. A 2% solution was prepared by dilution. Previous work by Wu <u>et al.</u> (26) has shown that this polymer has a number average molecular weight, M_n, of approximately 1.25×10^5 and a polydispersity index, M_w/M_n, of 3.0. A Headway Research spinner was modified to permit selection and control of the rate at which the final rotational speed is approached. Unless otherwise specified, a 1-sec. ramp was used for most runs. The vacuum chuck and spinner bowl were enclosed in a glove box, which was slowly vented by a vacuum line. This arrangement allows a fairly constant solvent vapor pressure to be maintained within the enclosure during spincoating. After spinning, the wafers were baked in a convection oven at 160°C for one hour to remove the residual solvent. This drying process reduced the film thickness by approximately 10%. Film thickness profiles were measured with an IBM 7840 Film Thickness Analyzer. This instrument uses an interferometric technique to determine the film thickness.

Experimental results and model predictions based on the non-Newtonian analysis are summarized on a log-log plot of average film thickness as a function of the final spin speed (Figure 2). Excellent agreement is seen between the model predictions and experimental data for 9,6 and 4% resist solutions. Serious discrepancy occurs, however, for the 2% polymer solution, the calculated thickness being significantly thicker than the experimental values. (Still, the slope of the predicted curve is in agreement with the data). The most likely source of this error is neglect of surface forces in the model. A 2% polymer solution spun at 1000 rpm leaves a final film that is approximately 200 nm thick, corresponding to only a few monolayers of polymer. The failure to model the spin coating of films from a 2% solution is not a serious problem since solutions with a polymer weight fraction below about 4% are generally not used in commercial practice. The reason for this is that very thin films often contain a large number of pinholes.

The significance of the non-Newtonian characteristics of the resist solution is illustrated in Figure 3. It is apparent that the assumption of Newtonian behavior leads to a weaker dependence of film thickness on spin speed than is observed experimentally, and one which is independent of polymer concentration in the initial resist. The non-Newtonian model not only gives quantitative predictions of the film thickness, but also generates the correct dependence on spinner speed. The dependence on spinner speed is stronger for the higher concentration solutions. This is attributable to the more prominant non-Newtonian behavior of the resist at higher concentrations.

It is of interest to examine the relative importance of convective and evaporative loss terms in film thickness

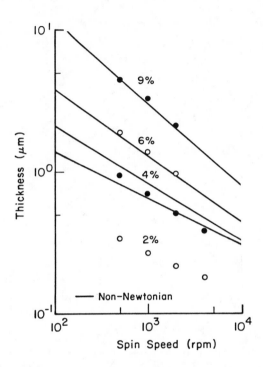

Figure 2. Comparison of measured and predicted average film thickness as a function of final spin speed for several resist concentrations. Increasing speed decreases film thickness, whereas increasing initial solution concentration leads to thicker deposits. Model predictions (solid lines) agree well with experimental observations (open and filled circles) for reasonably concentrated solutions. At low initial concentrations, e.g., 2% to 4%, the model overpredicts the final film thickness. (Reproduced with permission. Copyright 1984, Journal of Applied Physics.)

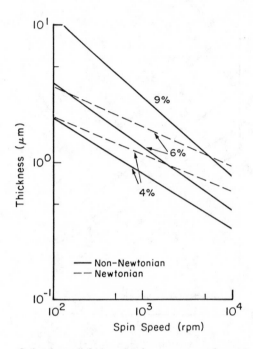

Figure 3.   Calculated film thickness as a function of spin
speed and concentration.   Solid curves represent the non-
Newtonian analysis, while dotted curves are for the Newtonian
model.   Neglect of shear-thinning behavior not only results in
overprediction of film thickness but also underprediction of
its dependence on spin speed.   (Reproduced with permission.
Copyright 1984, Journal of Applied Physics.)

reduction. Here, we choose the case of a 6% solution with an
initial thickness of 450 μm (equivalent to 3 ml liquid flooding a 3
in. wafer), spun at a final spin speed of 1000 rpm. Figure 4 gives
film thickness loss rates attributed to convective flow and solvent
evaporation as functions of time. Convection clearly dominates
during the early stages of the process, but evaporation takes over
at longer times. The convective loss curve exhibits a sharp rise
initially. This is due to spinner acceleration, which increases
the effectiveness of the spinner to throw off fluid radially. The
rapid decrease in the rate of convective loss following the maximum
is due to the combined effects of the increasing viscosity of the
resist solution and the increasing significance of the no-slip
boundary condition at the wafer surface. The rise in the rate of
evaporative loss is due totally to the decrease in film thickness. This is moderated, of course, by the rapidly decreasing
solvent diffusivity.

The seemingly straightforward gradual trade-off between the
two loss mechanisms seen in Figure 4 conceals the complex interplay
of various process variables. The loss mechanisms are closely
coupled, and the relationships among the process variables can be
subtle. Hence, the net effects of a change in one process variable
cannot be projected intuitively. For example, increasing solvent
partial pressure in the chamber would be expected to reduce solvent
evaporation and hence contribute to forming a thicker film. However, the slower drying process also retards the fluid viscosity
rise and thereby enhances convective loss. This latter process
contributes to forming a thinner film. It is therefore apparent
that in order to understand how solvent vapor pressure, or for that
matter any other variable, affects the final film thickness, it is
necessary to utilize the complete model.

Before leaving this phase of resist processing, it is worth
mentioning that detailed film thickness profiles have been found to
vary with spinner acceleration characteristics, even though the
same final spin speed is attained. This is an important problem
confounding practitioners who have observed different resist film
qualities from different spinners while the spin-coating process is
purportedly conducted under identical conditions (e.g., at the same
final speed). The non-Newtonian model successfully predicts the
observed dependence of thickness profile on spinner acceleration
rate (15). In addition, as demonstrated above, model predictions
also agree quantitatively with experimental observations of the
dependence of film thickness on concentration and final spin
speed. This model can therefore be used as a design guide or an
evaluation tool for new resist systems whose material properties
are known.

The success of modeling the spin-coating process achieved here
is only preliminary. Deposition of resists over an existing con-
toured topography is the next challenge, and a formidable one. The
complexity of this problem, i.e., fluid flow over a step protrusion
or cavity, can be appreciated by noting that the direction of radial flow may be at any given angle from the axis of the protrusion
or cavity. Reasonably realistic models will have to encompass flow
external to the cavity or over the protrusion, as well as the possible development of internal circulation within the cavity or near

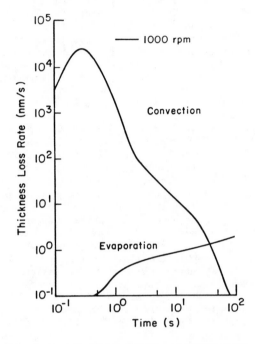

Figure 4. Calculated thickness loss attributed to convective motion and solvent evaporation for processes involving a 6% KTI-PMMA solution spun at 1000 rpm. Convection governs early phases, whereas evaporation contributes primarily to thickness reduction in later stages. The initial rise in convective loss is caused by spinner acceleration. (Reproduced with permission. Copyright 1984, Journal of Applied Physics.)

the base of a vertical protrusion. Successful modeling efforts are
expected to create a major impact on bi-level and tri-level resist
technology (27,28).

Exposure

Relevant engineering considerations of this aspect of the resist
technology are numerous. For example, the substrate and radiation
source must both be held rigidly. Advances in vibration isolation
and mechanical stability have been made in the past; these factors
no longer limit the process. Since the energy density from a
source, much like resist sensitivity, dictates the exposure time,
the radiation source must maintain both temporal and spatial
stability during exposure. Diffraction of UV often limits latent
pattern resolution, and precise wafer alignment is critical. Since
electron and ion-beam sources operate in vacuum, the charge
deposited in the substrate must be conducted away efficiently.
Finally, cleanliness of the exposure hardware to minimize
particulate contamination is of great importance to the yield.
Many other similar engineering considerations exist. We will,
however, not be concerned with these mechanical and hardware
considerations. Rather, we will focus on major events occurring in
the resist layer during exposure.

Resist imaging on highly reflective surfaces, such as
aluminum, gives rise to linewidth deviations, when a change of
reflectivity of the underlying surface is encountered. The wafer
reflectivity and topography both contribute to interference of the
incident light with the reflected light. For example, a 1 μm
resist coating over a 1 μm step may thin to 0.5 μm at the top of
the step and may be as thick as 1.5 μm at the bottom. Normally, a
thickness change of less than a quarter of the radiation wavelength
(by a factor equal to the refractive index of the resist) will
change the exposure energy coupling with the reflected light from
constructive interference (maximum local intensity) to destructive
interference (minimum local intensity). A number of such constructive and destructive couplings can then occur within the resist.
The latent exposure energy profile along an arbitrary vertical axis
thus exhibits multiple extrema. Upon development, edges of a step
may show a scalloped pattern as a result of the different rates of
dissolution corresponding to uneven exposure density.

In addition to this local variation, a global drift of the
scalloped edge profiles often accompanies the varying resist layer
thickness over a criss-crossing step beneath. This problem has
been researched by Neureuther et al. (29). A model has been
developed to simulate this exposure aberration. Figure 5 displays
calculated resist line edge profiles typically seen at step cros-
sings due to resist thickness change from a thin layer over the top
to a thick one near the bottom. Indeed, scalloped edge pattern is
prominent. The linewidth also drifts from that corresponding to
the top of a step to one over the bottom. This optical interference model constitutes an element of a larger scale effort
called SAMPLE, which is a coherent family of simulation programs
(30). The key objective of SAMPLE is to establish a single master
program simulating different fabrication processes in a user speci-

fied sequence. The growth of SAMPLE is attained by continual incorporation and refinement of individual process models such as this one. Resist processing is a vital module of this large-scale effort.

The above problem caused by scattered reflections over existing wafer surface topography is most pronounced with the use of direct wafer steppers using monochromatic exposing radiation. Recently, spun-on antireflective coatings have been shown to be effective in minimizing wafer surface reflections and improving resist performance (31).

An analogous problem challenging e-beam exposure of resists is electron back-scattering from the resist-substrate interface. This phenomenon gives rise to a greater energy distribution, thus more extensive structural damage of the resist, near the wafer-resist interface. Monte-Carlo programs tracing the paths of electrons through the resist layer and a silicon base have been written (32-34). Figure 6 shows a typical plot of simulated electron trajectories. Figure 7 gives a contour map of the corresponding energy distribution in the resist.

We wrap up the discussion of this phase of resist processing by mentioning that for x-ray and ion beam lithography, energy deposition models have been developed to simulate PMMA resist systems (35-37).

Development

The development process converts the latent image in the polymer into the final 3-D relief image. This process is perhaps the most complex of resist technology. It can generally be achieved by either liquid development or dry (plasma) development. Numerous considerations are critical to either alternative. We will first focus on the wet development process. Plasma development will be discussed in a later section.

Pattern fidelity must be maintained during polymer dissolution, which invariably exploits the existence of different rates of dissolution between the exposed and unexposed regions. Network swelling, inadequate contrast in rates of dissolution, and lack of reproducibility of the development conditions are major problems compromising the attainment of precise final patterns and linewidths. Swelling in regions adjacent to the irradiated area may be caused by residual stresses induced in the spin-coating process that have not been completely relieved in prebaking. It could also be simply due to solvent imbibition. Rates of polymer dissolution are strong functions of the molecular weight, polymer-solvent interaction (compatibility), network elastic stress, temperature, degree of agitation (which changes external mass transfer rates relative to solvent mobility in the polymer matrix). Hence, solution thermodynamic properties, kinetic factors such as solvent diffusivity in the resist, polymer dynamics determining coil unraveling from the swollen entanglement network, and stress growth and relaxation of the polymer matrix due to solvent penetration, all govern the apparent dissolution rates. These factors are amenable to more in-depth theoretical analyses, as our knowledge of polymer physics at present is sufficiently advanced to enable an

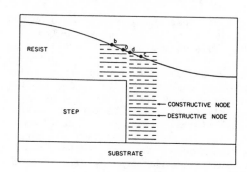

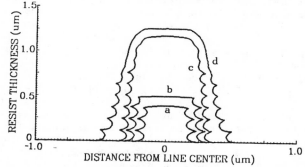

Figure 5. Exposure artifacts illustrated by the constructive and destructive interference nodes in the resist layer over an underlying step (top figure). The resulting scalloped line edge profiles over such a step crossing showing global linewidth drifts as resist layer thickness changes (bottom figure). (Reproduced with permission from Ref. 29. Copyright 1981, SPIE.)

100 20KV Electron Paths

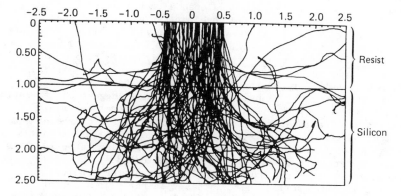

Figure 6. Monte-Carlo simulation of electron trajectories for a 20 kV source.

effective treatment of these phenomena. Tu and Ouano (38) have
described the dissolution process of glassy polymers by a pheno-
menological model. This work provides an excellent framework for
future incorporation of a more detailed description of the fundamental thermodynamics and kinetics accompanying resist development.

The dissolution of exposed resists in developer solutions is a
critical lithographic step in the fabrication of microelectronics.
A program was recently initiated at Berkeley to study this process
(39). The objective of this effort is to establish optimal process
conditions to maximize linewidth resolution and pattern fidelity,
while minimizing development time. Since these desired goals tend
to be mutually exclusive, trade-offs in performance must be
tolerated. A number of process parameters, e.g., developer
strength, temperature, and molecular weights in the exposed and
unexposed regions, significantly affect the outcome of the
process. Fundamental studies of the complex dependence of resist
development on these process variables will greatly facilitate
future process design and refinement.

Critical information needed for such fundamental understanding
and mathematical modeling of the resist development process entails
the determination of resist film thickness as a function of time in
the developer solution. Traditional techniques rely on repeated
development of a resist for various predetermined times, followed by
quenching in a nonsolvent and baking to drive off the developer
before measuring the remaining film thickness (40,41). This technique is prone to errors due to the presence of induction effects
and failure to stop dissolution immediately when desired. In-situ
monitoring of the process became a reality only recently. Konnerth
and Dill (42) used a computer controlled spectrophotometer to measure the relative reflectivity of the sample as a function of wavelength. Film optical properties and a complex method for converting
relative to absolute reflectivity were necessary to determine film
thickness. A laser endpoint detection system based on inter-
ferometry principles has been designed by Willson (43). This device
provides data to within an integer multiple of one-half a
wavelength. Oldham (44) has developed a technique for measuring the
capacitance between conductive developer solutions and the
substrates. The total thickness of dielectric is inversely related
to the capacitance. Recently, information from standing-wave shapes
of a HeNe laser system has allowed better than $\lambda/2$ resolution (45).

Our experimental program is based on a simplified ellipsometer
(psi meter) (46) for the in-situ measurement of resist film thickness in developer solutions. Preliminary data suggest the formation
of a gel layer over the remaining glassy film under certain
conditions. The interpretation of this observation can be discussed
in light of the relative rates of solvent penetration into a glassy
matrix and polymer coil detachment from the swollen network. This
will be deferred to a forthcoming publication (39). We will
concentrate only on the experimental aspect of the project.

The system described here is based on the principles of
ellipsometry (47). Figure 8 gives a block diagram of our experimental setup. The Psi-meter consists of a 2 mW helium-neon polarized laser source, a quarter wave retardation plate to generate
circularly polarized light, a synchronously rotating polarizer to

Energy Distribution Plot

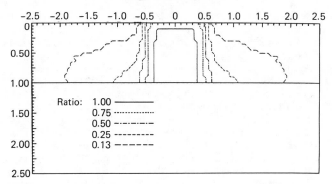

Figure 7.   The energy distribution contour map based on the Monte-Carlo simulation work depicted in Fig. 6.   Increasing electron back scattering probability leads to a substantially distorted energy deposition pattern.

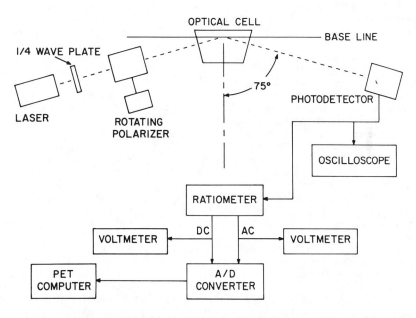

Figure 8.   Schematic block diagram of the Psi-meter setup with the major optical and detection components identified.   (Reproduced with permission from Ref. 29.   Copyright 1984, The Electrochemical Society.)

convert the beam to linearly polarized light with time-dependent
orientation. Light reflected from the wafer placed in the optical
cell enters a photodetector. The signal from the detector is split
into the AC and DC components by the ratiometer. Final data
acquisition and storage are achieved with a dedicated Commodore PET
2001 computer.

The amplitude ratio of the AC to DC components is equal to
$-\cos(2\psi)$, and ψ is related to the optical constants of the
substrate, film, and ambient, the wavelength of the light, incidence
angle and film thickness through the Drude equation (47).

Silicon wafers coated with approximately 700 nm PMMA resist and
prebaked at 160°C for one hour were used in this study. Film
dissolution took place in the optical cell, where a developer
solution at room temperature was circulated at 100 ml per minute
with the aid of an external pump. The optical cell has a capacity
of 250 ml. The motor-driven polarizer was adjusted to a rotation
frequency of 100 Hz. An incident angle of 75° was chosen to
maximize sensitivity. The wavelength was 632.8 nm. Data were
collected at a rate of one point per second, allowing detailed
delineation of the dissolution phenomenon.

Figure 9 shows plots of AC/DC amplitude ratios as a function of
dissolution time for wafers immersed in developer solutions of
methyl ethyl ketone (MEK) and isopropanol (IPA) with different
compositions. Appreciably different time scales are needed to
achieve dissolution, with shorter development times for solutions of
greater strengths (higher concentrations of MEK, the good solvent). The curve for the stronger developer solution, after decon-
volution by a program attributed to McCrackin (48), indicates a
fairly linear decrease in film thickness with time, with no appreciable induction period and no distinct multiple layer formation.
An extra maximum, however, emerges in the AC/DC at short times for
the weaker solution. This intriguing observation leads to several
possible hypotheses to rationalize its existence. At present, we
suspect that this extra maximum is associated with the formation of
a gel layer at the solution-resist interface. Polymer dissolution
involves first solvent penetration into the glassy matrix, converting it into a rubbery gel phase, followed by detachment of chain
molecules from the entangled gel network into the solution. The
former process is influenced by the mobility of small solvent
molecules penetrating the glassy phase, whereas the latter is
affected by chain diffusion. A thorough analysis of the resist
dissolution kinetics will further elucidate the underlying physics.

An independent observation of surface induction on an entirely
different material has been made by Kim et. al. (45) using an
interferometric technique. Kodak 820 photo-resist-spun on a glass
substrate is dissolved by two developers. The interference intensity is tracked as a function of time, allowing the development rate
at every 0.01µm to be resolved. Figure 10 gives a typical plot,
showing that film dissolution is slowest near the surface (even
though the local exposure is highest). This sort of surface
induction has also been reported previously (49).

The phenomenon of surface induction could be interpreted with
either the concept of gel-layer formation or a surface hardening
effect due to baking. When the glassy matrix is first brought in

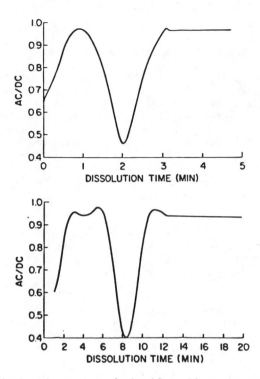

Figure 9. Amplitude ratios of the AC to DC components as functions of time for PMMA dissolution in MEK and IPA developer solutions with compositions (volume ratios of MEK to IPA) of 6:4 and 4:6. (Reproduced with permission from Ref. 29. Copyright 1984, The Electrochemical Society.)

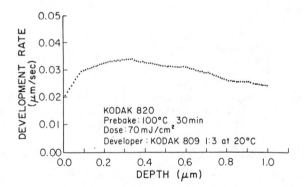

Figure 10. Development rate vs. depth for KODAK 820 resist developed in 809 type developer. The surface layer dissolves at a lower rate than other parts of the resist, indicating a surface induction effect. (Adapted from Ref. 45.)

contact with the developer solution, small molecule penetration
ensues. Polymer disengagement from the entwined network does not
occur instantaneously and only becomes significant when the top
layer of the film is sufficiently swollen. The degree of this
critical swollen state is a function of various parameters,
including polymer-solvent compatibility, solvent diffusivity, and
polymer coil size, etc. A thorough analysis is needed. An
alternate explanation of surface induction is hardening of resist
top layer due to excessive drying or chemical modification such as
crosslinking. The effects of surface induction and formation of
multilayer structure on the final edge shape and linewidth distortion may be extensive. Finite element analyses and experimental
verification of model predictions in ths area constitute a viable
research program.

Baking

Engineering considerations are similar in many respects for
prebaking and post-baking. The polymer must achieve a uniform,
desired temperature within a short time and be maintained at the
temperature for a sufficient duration. This temperature must be
high enough to remove residual solvent and locked-in stresses, yet
not so high as to cause degradation of sensitizers. Melt flow must
be prevented in post-bake, yet can be used advantageously to
smoothen uneven topography during pre-bake. Resist structure
alteration by thermal degradation must also be avoided. Finally,
cleanliness of sample handling is critical.
      Fundamental research opportunities in this processing step
remain, despite its apparent simplicity. As the solvent leaves the
resist, the free volume fraction of the matrix decreases, significantly slowing further relaxation processes. Complete polymer
relaxation is necessary as it not only alleviates stresses induced
in spin-coating, but also contributes to increased density. Film
consolidation is hampered by the reduced free volume associated with
solvent molecules as the material dries, unless the system temperature is much higher than T_g of the polymer. A priori selection of
the optimal baking temperature and duration is not as straightforward a task as it seems at a first glance. Yet, the state of the
resist after baking is critical in affecting the following steps.
For example, developer penetration into consolidated polymer is
slower than a matrix containing residual solvent. Also, stress
relief during development can cause pattern distortion. Bire-
fringence or wafer bowing phenomenon may be used as possible tools
to probe the stress state of resist films.

Plasma Development and Etching

Plasma development derives its advantage over its liquid counterpart
mainly from the anisotropic nature of the process, except when loss
from the unexposed area is vanishingly small, in which case the
isotropy comes from that already in the latent image. The resist
must be tough to avoid being eroded completely while the substrate
underneath is etched. To improve the plasma etch resistance,
aromatic compounds have been added to PMMA (50, 51). Hence, for

etch protection, the resist cannot be sensitive to degradation. However, for a positive resist, the macromolecular structure must degrade easily upon irradiation (by UV, x-ray, or e-beam) to optimize the sensitivity during exposure. This material requirement is in direct conflict with that of plasma etching. A resist material is thus required to display two incompatible characteristics: high radiation sensitivity and low plasma degradation. The apparent inconsistency in resist property optimization and its ramifications beg for research and development efforts. Satisfactory solutions only exist for several well-researched systems.

We have carried out a study on the degradation of PMMA exposed to CF_4 and CF_4/O_2 plasmas in a parallel plate reactor by means of gel permeation chromatography (26). Since a plasma (glow discharge) is an extremely complex mixture of ions, electrons, photons, atoms, and free radicals in ground and in excited states, exposure of film materials to this environment results in ion, electron, and photon bombardment, promoting various chemical processes on the surfaces. Further, substantial heating of the resist (along with the substrates) can occur during plasma processes. Thus, many resist materials erode and degrade rapidly during plasma etching.

The penetration depth of plasma interaction with a polymer has been estimated to vary between 50Å and 10 μm (52). The specific depth of interaction (crosslinking, degradation, chemical modification) depends on both the polymer and the plasma conditions (power, pressure, etc.). Therefore, it would be useful to ascertain the depth and mode of plasma interaction during etching, since the resist films are typically a micron or less in thickness. If, for example, the degradation effects are confined to the top few hundred Angstroms of a film, a surface hardening treatment may be all that is required to protect the resist.

Comparison of the molecular weight distribution before and after plasma etch gives a direct indication of the extent of degradation. Gel permeation chromatography (GPC) is a convenient method of analyzing molecular weight distributions (53).

Harada has used GPC to profile plasma degradation in PMMA films (54). Wafers with different initial resist thicknesses were simultaneously etched with CF_4/O_2 in a barrel reactor. The residual resists were then completely stripped and analyzed by GPC. Harada calculated an overall resist degradation profile by subtracting the molecular weight distributions of successive thickness of PMMA. However, in so doing, he implicitly assumed that the molecular weight distributions are similar for successive layers. In general, this assumption is not valid. In our study, the PMMA degradation was profiled by dissolving successive layers of the same resist film, and then analyzing each layer for molecular weight distribution by GPC.

Major results of this work are summarized in Figure 11, which shows a plot of the number average molecular weight versus resist depth computed from the GPC data. Note that reduction of molecular weight decreases with depth in the film, but degradation occurs even at the deepest levels shown. The degradation profiles for CF_4 and CF_4/O_2 exposed films are virtually identical. However, the superficial etch rates of the CF_4 and CF_4/O_2 treated films were found to be 100 and 500 Å/min., respectively. This suggests that

the effect of O_2 may be confined to the surface of the resists,
thereby merely enhancing the chemical etch rate. It may be more
generally inferred that the superficial etch rate and the degradation (molecular weight loss) within the film do not necessarily
correlate with each other.

The relative rate of molecular weight degradation is shown in
Figure 12. The effects of the plasma are particularly intense
within the top few thousand Angstroms of the film. The rapid
decrease in degradation with depth can be explained by either the
shallow penetration depth of ions or electrons that bombard the
resist surface or by the rapid consumption of reactive species (e.g.
halogen radicals) which diffuse into the polymer matrix.

The steep drop in degradation rate with depth also suggests
that any photochemical contribution (except for deep UV, perhaps,
where absorption can be strong) is relatively small. The degradation profile would have been more uniform had UV degradation
predominated because of the generally low optical absorption of PMMA
(53). Optical emission studies of CF_4 and CF_4/O_2 plasmas have shown
that there are relatively few emission lines in the 200-300 nm
wavelength region (55). The possibility of deep UV induced degradation cannot be ruled out entirely. However, the probability for
the two plasmas to emit deep UV at the same intensity is small. Yet
PMMA molecular weight profiles produced under both conditions are
nearly identical.

The above observations are in general agreement with a random
scission model. However, it is presently unclear whether random
scission is the predominant mechanism at the plasma-resist interface. Some crosslinking and/or reaction with fluorine-containing
fragments or oxygen radicals may take place as well. The experimental techniques used in our work to isolate degraded PMMA in successive layers did not have sufficient resolution to permit analysis
of the top one or two thousand Angstroms of resist. Analysis of
plasma degradation of surface-modified PMMA with GPC techniques
represents a possible extension, designed to elucidate events
occurring at the plasma-resist interface. Study of the species in
the discharge as well as the effluent also constitutes major
research opportunities.

Conclusion

A brief overview is devoted to several current research projects in
resist processing. The problems under study cover all the important
processing steps, including spincoating, exposure, development,
baking, and plasma development and etching. Approaches adopted to
gain a better understanding of the governing processes for each step
draw on fundamental knowledge of polymer physics, mass and energy
conservation, kinetic aspects of transport phenomena, and dynamics
of macromolecules. The discussions in this paper reflect the per-
spectives of a chemical engineer with a traditional polymer research
background. Numerous other potential research areas can undoubtedly
be identified. This paper also neglects the chemistry and chemical
synthesis aspects of resist materials. Attention has only been paid
to one-dimensional processes and single-layer technology. Resist
planarization and multi-layer resist systems (56) have not been the

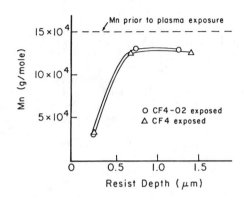

Figure 11. Number average molecular weight of PMMA vs. resist depth after plasma exposure. The points represent the midpoints of the 0.5 m film segments sampled. (Reproduced with permission from Ref. 26. Copyright 1983, Journal of Applied Physics.)

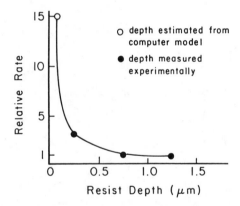

Figure 12. Relative rate of resist degradation vs. resist depth. The rate near the surface (open circle) was estimated by matching the molecular weight distribution profile predicted by a random scisson model with that measured by GPC. Again, the points represent midpoints of the sampled film segments. (Reproduced with permission from Ref. 26. Copyright 1983, Journal of Applied Physics.)

main focus of the present paper. However, a good grasp of the fundamental processes governing single-layer technology lays the foundation for future extension into multi-level processes.

Acknowledgment

The author very much appreciates the continuing collaboration with and support from his colleagues, Dennis W. Hess and Alexis T. Bell. Former and present graduate students, Bernard J. Wu, Lee M. Gavens, Warren W. Flack, James S. Papanu, and Joseph Manjkow have made resist technology an exciting area for the author, who was originally interested in traditional polymer processes. Information supplied by Professor Andrew R. Neureuther on the SAMPLE program at Berkeley is also greatly appreciated. Our research has been funded by the Air Force Office of Scientific Research Under Grant AFOSR-80-0078.

Literature Cited

1. Tai, K. L., Sinclair, W. R., Vadimsky, R. C., and Moran J. M. J. Vac. Sci. Technol., 1977, 16.
2. Willson, C. G, in "Introduction to Microlithography," ACS Symp. Series, 1983, 219, 86.
3. Thompson, L. F. and Bowden, M. J. in "Introduction to Microlithography," ACS Symp. Series, 1983, 219, 162.
4. Emslie, A. G., Bonner, F. T., and Peck, L. G., J. Appl. Phys., 1958, 29, 858.
5. Acrivos, A., Shah, M. J., and Petersen, E.E., J. Appl. Phys., 1960, 31, 963.
6. Dorfman, L.A., J. Eng. Phys, 1967, 12, 162.
7. Damon, G. F., Proc. Kodak Seminar on Microminiaturization, 1969, 34, 195.
8. Fraidenraich, N., Revista Mexicana de Fisica, 1976, 25, 69.
9. Washo, B. D., IBM J. Res. Develop., 1977, 21, 190.
10. O'Hagan, P. O., and Daughton, W. J., Proc. Kodak Seminar on Microelectronics, 1977, G-48, 95.
11. Meyerhofer, D., J. Appl. Phys., 1978, 49, 3393.
12. Daughton, W. J., O'Hagan, P. O., and Givens, F. L., Proc. Kodak Seminar on Microelectronics, 1978, G-49, 15.
13. Givens, F. L., and Daughton, W. J., J. Electrochem. Soc., 1979, 126, 269.
14. Daughton, W. J., and Givens, F. L., J. Electrochem. Soc., 1982, 129, 173.
15. Flack, W. W., Soong, D. S., Bell, A. T., and Hess, D. W., to be published in J. Appl. Phys.
16. Baillagou, P. E., and Soong, D. S., to appear in Chem. Eng. Comm.
17. Doolittle, A. K., J. Appl. Phys., 1951, 22, 1471.
18. Fujita, H., and Kishimoto, A., J. Chem. Phys., 1961, 34, 393.
19. de Gennes, P. G., "Scaling Concepts in Polymer Physics," Cornell University Press, Ithaca, N.Y., 1979.
20. Williams, M. L., Landel, R. F., and Ferry, J. D., J. Am. Chem. Soc., 1955, 77, 3701.
21. Soong, D. S., and Shen, M., J. Rheol., 1981, 25, 259.

22. Liu, T. Y., Soong, D. S., and Williams, M. C., Polym. Eng. Sci., 1981, 21, 675.
23. Liu, T. Y., Soong, D. S., and Williams, M. C., J. Rheology, 1983, 27, 7.
24. Vrentas, V. S., and Duda, J. L., J. Polym. Sci., 1977, 15, 417.
25. Vrentas, V. S., and Duda, J. L., AIChE J., 1979, 25, 1.
26. Wu, B.J., Hess, D. W., Soong, D. S., and Bell, Bell, A. T., J. Appl. Phys., 1983, 54, 1725.
27. Lin, B. J., and Chaing, T. H. P., J. Vac. Sci. Technol., 1979, 16, 1669.
28. Lin, B. J., Bassous, E., Chao, V. W., and Petrillo, K. E., J. Vac. Sci. Technol., 1981, 19, 1313.
29. Neureuther, A. R., Jain, P. K., and Oldham, W. G., SPIE, 1981, 275, 110.
30. SAMPLE reports and related projects at UC Berkeley, 1983 update, Electronics Research Laboratory, University of California, Berkeley.
31. Coyne, R. D., and Brewer, T., Kodak Microelectronics Seminar, November 1983.
32. Kyser, D. F., and Pyle, R., IBM J. Res. Dev., 1980, 24, 426.
33. Murata, K., Kyser, D. F., and Ting, C. H., J. Appl. Phys., 1981, 52, 4396.
34. Neureuther, A. R., Kyser, D. F., and Ting, C. H., IEE Trans. Electron Devices, 1979, ED26, 686.
35. Neureuther, A. R., J. Vac. Sci. Technol., 1978, 15, 1004.
36. Heinrich, K., Betz, H., Heuberger, A., and Pongraz, S., J. Vac. sci. Technol., 1981, 19, 1254.
37. Karapiperis, L., Adesida, I., Lee, C. A., and Wolf, E. D., J. Vac. Sci. Technol., 1981, 19, 1259.
38. Tu, Y. O., and Ouano, A. C., IBM J. Res. Dev., 1977, 21:2, 131.
39. Flack, W. W., Papanu, J. S., Hess, D. W., Soong, D. S., and Bell, A. T., preliminary communication to appear in J. Electrochem. Soc., full manuscript under preparation.
40. Dill, F., IEEE Trans. Electron. Dev., 1975, 22, 445.
41. Ouano, A. C., Polym. Eng. Sci. 1978, 18, 306.
42. Konnerth, K., and Dill, F., IEEE Trans. Electron Dev., 1975, 22, 452.
43. Willson, G., IBM Corporation, San Jose, CA, private communication.
44. Oldham, W., SPIE Semiconductor Microlithography III, 1978, 135, 153.
45. Kim, D. J., Oldham, W. G., and Neureuther, A. R., Proceedings Kodak Seminar on Microelectronics, October, 1983.
46. Zaghloul, A., and Azzam, R., Surface Sci., 1980, 96, 169.
47. Azzam, R., and Bashara, N., "Ellipsometry and Polarized Light," North Holland Publishing co., New York, 1977.
48. McCrackin, F., "A Fortran Program for Analysis of Ellipsometer Measurements," National Bureau of Standards, 1969, Technical Note 479.
49. Dill, F. H., Hornberger, W. P., Hauge, P. S., and Shaw, J. M., IEEE Trans. Electron Devices, 1975, ED22:7, 445.
50. van Pelt, P., SPIE, 1981, 275, 150.

51. Taylor, G. N., and Wolf, T. M., Proc. S.P.E. Conference on
     Photopolymers, 1979,p. 174, Ellenville, N.Y., October, 1979.
52. Coopes, I., and Grifkins, K., J. Macromol. Sci. Chem., 1982,
     A17, 217.
53. Chandross, E., Reichmanis, E., Wilkins, C., and Hartless, R.,
     Solid State Technol., August, 1981, p. 81.
54. Harada, K., J. Appl. Polym. Sci., 1981, 26, 1961.
55. Harshbarger, W., and Porter, R., Solid State Techniol., April
     1978, p.99.
56. Lin, B. J., in "Introduction Microlithography," ACS Symp.
     Series, 1983, 219, 287.

RECEIVED August 1, 1985

Physical and Chemical Modifications of Photoresists

Peter C. Sukanek

Department of Chemical Engineering, Clarkson University, Potsdam, NY 13676

The physical changes which occur in a resist during spin coating and the chemical changes which may occur during various image stabilization techniques are discussed. To predict the film thickness in spin coating, macroscopic mass balance equations are written for a resist system which contains a nonvolatile component and a volatile component which evaporates during the spinning. The model predictions are in qualitative agreement with experimental data. The film height is always proportional to the inverse square root of the spin speed. The factors which may influence resist image deformation at high temperature are also reviewed. Some of the proposed stabilization methods and the mechanisms by which they are thought to work are presented. Unknown factors in both coating and stabilization are highlighted.

This paper investigates two problems frequently encountered in the processing of resist materials. These are, first, the physical transformation of the resist from a viscous liquid to a solid-like film on the substrate (usually a wafer) surface. The second problem is how this film is chemically altered during the various image stabilization steps suggested in the literature. What often happens in this step is that the resist is transformed from a soluble material to an insoluble one. The extent of the image deformation on heating probably depends on the properties of the film, specifically the softening point (and hence the molecular weight) and the extent of retained solvent after spinning. Despite the widespread use of both of these processes in the industry today, several unknowns remain.

Spin Coating: Development of the Equations

The first published analysis of spin coating is due to Emslie, Bonner and Peck (1), who treated the problem of coating a single component, non-volatile, newtonian fluid on a disk. Film height

0097–6156/85/0290–0095$06.00/0

continuously decreases with time as the fluid is "pumped" by centrifugal force along the disk. Since the viscosity of the fluid is assumed constant, this pumping mechanism is always present, and the film thickness eventually approaches zero. Acrivos, Shah and Petersen (2) later treated the problem of coating a power-law fluid. Unlike the newtonian result, a uniform film thickness cannot be obtained in this case. A layer which is initially uniform becomes non-uniform as a result of the spinning. The more the departure of the fluid's behavior from newtonian, the more non-uniform the film. Again, the film height approaches zero at long times. In order to achieve a finite film height, a constant source of fluid must be supplied at the center of the disk. Whereas the above two investigators used as the starting point in their analyses macroscopic mass balance equations, Washo (3) attempted to utilize the full Navier-Stokes equations from the start. Unfortunately, an error in his simplification of these equations invalidates his theoretical results. (Ignoring the small but important vertical velocity leads to an incorrect mass balance.)

Meyerhofer (4) examines the problem of spinning a two component system which is composed of "solids" and a volatile "solvent". He includes the effect of solvent evaporation and a viscosity which depends on solvent concentration. In this model, the film height decreases with time not only because of the centrifugal pumping, but also because of the evaporation from the surface. As the film is depleted in solvent, the viscosity increases and less fluid is lost because of pumping. Eventually a finite film thickness can be reached because all of the solvent evaporates, and the film consists of only an immobile solid phase. The present analysis is similar to Meyerhofer's. His treatment of the evaporation is corrected, as is his conclusion on the effect of spin speed (which apparently is due to numerical error on his part), and some of his assumptions in developing the equations are relaxed.

The fluid is assumed to consist of two components, a volatile component with concentration C_V and a nonvolatile component with concentration C_N. Mass balances on a cylindrical control volume on the disk yield:

$$\frac{\partial}{\partial t} (C_N h) = -\frac{1}{r} \frac{\partial}{\partial r} (r C_N q) \tag{1}$$

$$\frac{\partial}{\partial t} (C_V h) = -\frac{1}{r} \frac{\partial}{\partial r} (r C_V q) - e \tag{2}$$

The flow rate, q, is given by (1, 2):

$$q = r\Omega^2 h^3 / (3\nu) \tag{3}$$

The evaporation rate, e, can be expressed in terms of a mass transfer coefficient, k:

$$e = k(C_V^* - C_{V\infty}) \tag{4}$$

where C_V^* is the concentration of the volatile component in the gas phase in equilibrium with the film, and $C_{V\infty}$ is the concentration in the gas far away from the disk. In most cases, this latter value can be taken as zero. If the distribution coefficient (or equilibrium constant) K is introduced, equation (4) becomes:

$$e = kKC_V \tag{5}$$

Finally, provided the air flow above the disk is laminar (i.e., the air Reynolds number is less than about 2×10^5) the Sherwood number varies with the square root of the air Reynolds number, and the mass transfer coefficient, k, is given by ($\underline{5}$):

$$k = a\Omega^{1/2} \tag{6}$$

Two final assumptions concerning the system are made. First, we look for a solution which depends only on time; that is, the height does not vary with radial position on the disk. Second, we assume that the density of the film is a constant and is independent of the concentration of either component.

With these assumptions, equations (1-3, 5 and 6) can be combined, rearranged and put in dimensionless form to give:

$$\frac{\partial y}{\partial \tau} = -\frac{2}{3} \frac{y^3}{\nu/\nu_o} - \alpha x \tag{7}$$

$$\frac{\partial x}{\partial \tau} = \alpha x(x-1)/y \tag{8}$$

where the dimensionless parameters are defined by:

$$y = h\Omega^{1/2}/(a^{1/2}\nu_o^{1/4}) \tag{9}$$

$$x = C_V/\rho \tag{10}$$

$$\tau = t\Omega a/\nu_o^{1/2} \tag{11}$$

$$\alpha = \nu_o^{1/4}K/a^{1/2} \tag{12}$$

Note that with these definitions, the spin speed is removed as a parameter from the defining equations.

The kinematic viscosity ratio is assumed to be of the same form as that used by Meyerhofer (4):

$$\nu/\nu_o = \nu_s/\nu_o + (1-\nu_s/\nu_o)\left(\frac{1-x}{1-x_o}\right)^n \tag{13}$$

Equations (7), (8) and (13) are solved using a Runge-Kutta integration. At sufficiently large values of τ, the dimensionless height, y, approaches a constant. From the definition of y, equation (9), the final height is given by:

$$h_f = y_f a^{1/2}\nu_o^{1/4}\Omega^{-1/2} \tag{14}$$

Hence, the height depends on the inverse square root of the spin speed. Further, provided the dependence on is not strong, the height also depends on the one-fourth power of the initial kinematic viscosity. These conclusions are in good agreement with experimental data.

Spin Coating: Results and Discussion

It should be noted that the governing equations do not account for any differences between static and dynamic dispense methods. Neither is the spinner acceleration included. While there is no simple way to incorporate these effects in this model, the experimental investigation of Daughton and Givens (6) shows that the final height is independent of the dispense speed over a range of 0 to 500 rpm and over an initial viscosity range of 28 to 1100 cp. Also, they find that the final height is independent of the acceleration from dispense speed to final spin speed in the range of accelerations of 3000 to 30000 rpm/sec. These values are typical of modern spin coaters.

The initial height of the fluid on the disk is unknown and does

not appear in this analysis other than as the initial condition for
the integration. Figure 1 shows the effect of this initial height,
y_0, on the final film height. Above a critical value of about 0.1,
the final height is independent of the initial value. This result
is in agreement with the experiments reported in (6), which show no
effect of the dispense volume on y_f. Using representative values
for the parameters, y_0 is about 0.2 for dispense volumes of 1 cm^3 on
a three inch wafer or 2 cm^3 on a four inch wafer. This assumes, of
course, that the fluid once dispensed uniformly covers the entire
disk. While this does not happen (typically, a dispensed
photoresist forms a puddle on the wafer surface which extends to
about 1 to 2 cm from the wafer edge), photographs in (6) show that
the wafer is fully covered in less than two revolutions.

Figure 2 shows that as the evaporation coefficient, α,
increases, the film height also increases. This is in qualitative
agreement with the work of Chen (7) on the effects of different
solvents on the final thickness of spun-on polymer films. It is not
possible to obtain a quantitative comparison with Chen's experiments
because of the empirical nature of his evaporation term. The more
volatile the solvent, the faster the viscosity of the fluid
increases, the less important the centrifugal pumping effect and the
greater the film height. A similar effect is found with increasing
the initial solids content, as shown in Figure 3. Adjusting the
initial solids loading by the addition of a thinner to the resist is
a common way to control film height. Figure 4 shows the effect of
the viscosity exponent, n, on the final height. As reported (4)
this parameter has little effect.

The results thus far have assumed that all of the volatile
component is removed during the spinning process. However,
according to the measurements of Dill, et al. (8,9) the film after
spinning can contain significant amounts of solvent. These
investigators showed that 20% of the weight of AZ-1350J resist film
after spinning is cellosolve acetate solvent. The effect of
residual solvent on final height is shown in Figure 5. The
evaporation coefficient was modified for these calculations as:

$$\alpha = \alpha'(x - x_R) \qquad (15)$$

where x_R is the mass fraction of residual solvent. As can be seen
from the Figure, residual solvent causes the final film height to
increase, but not linearly. That is, a film with 10% residual
solvent is not 10% thicker than one with no solvent. The increased
solvent content decreases the viscosity causing more fluid to be
pumped off the wafer by centrifugal action.

A number of experimental investigations into spin coating have
appeared in the literature. Different investigators correlate the
thickness with different parameters. Daughton and Givens (6) and
Chen (7) show a dependence on the initial kinematic viscosity to the
0.29 and 0.36 powers, respectively. This is in rough agreement with

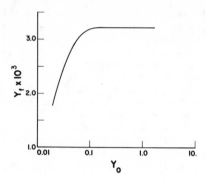

Figure 1. Effect of initial film height, y_0, on final height, y_f.
($x_0=0.9$;$\nu_s/\nu_0=10^{-2}$;$n=4$; $\alpha=10^{-5}$)

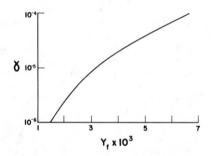

Figure 2. Effect of evaporation coefficient, α, on final film
height, y_f. ($x_0=0.9$; $\nu_s/\nu_0=10^{-2}$; $n=4$; $y_0=0.2$)

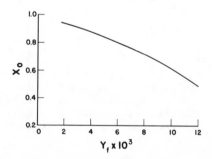

Figure 3. Effect of initial solids content, x_0, on final film
height, y_f. ($y_0=0.2$; $\nu_s/\nu_0=10^{-2}$; $n=4$; $\alpha=10^{-5}$)

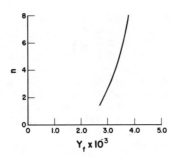

Figure 4.   Effect of viscosity exponent, n, on final film height, y_f. (y_o=0.2; ν_s/ν_o=10^{-2}; α=10^{-5}; x_o=0.9)

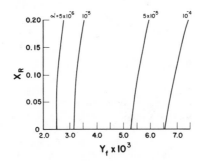

Figure 5.   Effect of residual solvent, x_R, on final film height, y_f. (y_0=0.2; ν_s/ν_o=10^{-2}; n=4; x_0=0.9; $\alpha = \alpha'(x-x_R)$)

the 0.25 power illustrated by equation (14). Others (10,11) show a
more   complex   dependence.   This,   however,   is   not   surprising
considering that initial viscosity and solids content usually cannot
be   varied   independently   and that the evaporation coefficient also
depends on initial viscosity.

Spin speed is the most common parameter used in the correlation
of film height, with most results (3,4,6,7,10,11) showing the
inverse square root dependence predicted by equation (14).   The
major discrepancy is with the results of Daughton and Givens (6) for
polyamic acid solutions, where a -0.81 power dependence was found.
While   it   is   tempting   to   attribute   this   result   to   possible
non-newtonian behavior of the solution, rheological measurements for
similar solutions (12) do not support this conclusion. The maximum
value of the shear rate (or the velocity gradient), γ , in spin
coating occurs at the interface between the wafer and the fluid.
For a newtonian fluid, the shear rate at this location is:

$$\dot{\gamma} = 3q/h^2 = \rho r\Omega^2 h/\mu \qquad (16)$$

For a 5µm thick film spun at 4500 rpm at the edge of a 3 in. wafer:

$$\dot{\gamma} \stackrel{\sim}{=} 40/\mu \qquad (17)$$

with viscosity, μ, in Pa s.

From Jenekhe's data (12) on what appears to be the same
polyamic acid solution as that used in (6), the newtonian region (μ
= 1.4 Pa s) persists out to a shear rate of about $300s^{-1}$.   This
indicates the flow is always in the newtonian region for the
experiments reported in (6).   It is, of course, difficult to
estimate the boundary between newtonian and non-newtonian effects in
this flow since as the solids content of the resist increases,
non-newtonian behavior may be expected to set in at a lower shear
rate. However, from equation (17), the shear rate experienced by
the fluid decreases as well due to the increased viscosity. In
order to resolve this question, more data is needed on the variation
of the rheological behavior with solids content.

The expression for the evaporation rate used in equation (4) is
based on laminar air flow. As the size of wafers used in the
industry increases from the most common 125mm size used today to the
relatively new 150mm and then to the 175mm and 200mm values recently
obtained by crystal growers, the air Reynolds number increases.
Using a typical value for the kinematic viscosity of air (15 mm^2/s),
the product of the rotation rate, ω, in rpm and the square of the
radius in mm must satisfy the following inequality for the air flow
to remain laminar:

$$\omega r^2 \gtrsim 1.43 \times 10^7 \qquad (18)$$

For a 125 mm wafer, the air flow at the very edge is no longer laminar once the rotation rate exceeds about 3700 rpm. At 6000 rpm, the transition occurs at a radial position of 49 mm; the outer 14 mm is in the transition or turbulent flow regime on a 125 mm wafer at this speed. In the turbulent region, the evaporation rate depends on radial position and varies with the spin speed to a power other than one-half. The result is expected to be a non-uniform resist thickness. While this effect is probably not important for the wafers in use today, it may become important in the future. The critical speed for a 200 mm wafer is only 1430 rpm. Above this value, at least part of the wafer will be in the turbulent air flow regime.

Image Stabilization: Methods and Mechanisms

In some processes, such as ion implantation and reactive ion etching, the resist can experience high temperature. When heated to temperatures above about $130^{\circ}C$, the resist can "reflow", that is, images present in the resist can deform. The surface forces which act on the film can cause the resist to flow into previously open areas and to pull back from the edges at other locations. In either event, the images are no longer satisfactory. The forces acting on the resist structures are complex, and it is not the intent of this paper to address the causes of this deformation. The "softening point" of typical resins used in resist lie in the range of 90-120$^{\circ}C$ (13). (The softening point is not a well-defined value.) Solvent still present in the resist after the spinning operation can act as a plasticizer, further reducing the softening point and leading to greater image distortion.

Several methods have been proposed in the literature to avoid or reduce the extent of the deformation. Certain resist systems are specifically marketed based on their stability to high temperature bakes. Other methods which may improve stability include slow heating, vacuum exposure with ultraviolet (UV) radiation, deep UV (at a wavelength of about 250 nm) exposure and exposure of the imaged resist to a plasma. In the remainder of this paper these methods will be discussed with the aim of gaining some insight into the chemical changes which occur in the resist during these operations.

In order to understand the changes, it is necessary first to know the components of the resist and the chemistry of the exposure step. For positive photoresists, the mechanism was elucidated by Pacansky and Lyerla (14). Typical resists are a solution of a naphthoquinone diazide photoactive compound (PAC) and a cresol formaldehyde novolak resin in one or more high boiling point industrial solvents. The number average molecular weight of the resin is quite low, on the order of 1000, with a polydispersity of about 10 (13). During the normal exposure step, the PAC, in the presence of water, absorbs light and is transformed into a

carboxylic acid. When water is absent the PAC reacts with the resin
to form an ester. If the PAC contains more than one active group, a
crosslinked polyester can result. This is the explanation of Tracy
and Mattox (15) for their vacuum exposure stability method. The
crosslinked structure is too rigid to undergo deformation at high
temperature.

A similar mechanism can be invoked to explain increased
stability as the result of vacuum or programmed (slow) heating (16,
17). In this case, however, two competing phenomena occur. As
demonstrated by Allen, et al. (18), when the resist is heated to
high temperature, it becomes insoluble in common solvents. However,
if the PAC is first deactivated, the resist is still soluble after a
high temperature bake. In addition images present in the resist
undergo greater distortion at high temperature if the PAC is
deactivated prior to bake. This suggests that a crosslinking
reaction similar to that discussed above occurs at high
temperature. In contrast with the vacuum exposure, during the high
temperature bake both image distortion and resist reaction occur
simultaneously. At higher temperatures the deformation apparently
occurs more rapidly than the chemical reaction, giving a distorted
image in an insoluble resist. If the resist is slowly heated, the
deformation rate is slower, and the chemical reaction causes some
stability of the images.

The explanation for stability after deep UV exposure is not
apparent, despite the fact that commercial instruments are available
to effect this stability. Different groups have examined resist
changes brought about by deep UV exposure(17-20), and it is
convenient to examine their results with respect to the wavelength
of light used, whether or not water is necessary, whether or not the
PAC is required and the resist type.

There is little agreement on which factors are important in
deep UV image stabilization. Hiraoka and Pacansky (17, 19) claim
that any wavelength is effective in achieving a stable image,
provided the PAC is deactivated. On the other hand, Allen, et al.
(18) find that this blanket exposure increases the deformation.
Vollenbroek, et al. (20) claim that the same mechanism is at work
here as in the vacuum exposure method; i.e., the PAC undergoes a
photochemical reaction with the resin to form a crosslinked
polyester. Therefore, their exposure must be done in the absence of
water. The use of 254 nm radiation confines the crosslinks to the
top surface of the resist, since the resin absorbs strongly at this
wavelength. Allen, et al. (18) were able to show that image
stability is linked to the formation of a "skin" on the surface of
the resist that results from the deep UV treatment. If 200 nm of
the resist surface is removed in a plasma, the images will again
deform. They exposed both the resin alone and resist with the
deactivated PAC to 254 nm radiation and found a surface skin in both
cases, suggesting that the reaction involves the resin only. It
seems clear that the deep UV radiation at 254 nm gives the resist a
"case" which contains it during subsequent high temperature bakes

and allows the thermal reaction to occur without the thermal deformation.

One final method of stabilization which has been proposed is to expose the resist to a glow discharge from a nitrogen or other gas plasma (21). Hansen and Schonhorn (22, 23) found that polyethylene and other polymers underwent a surface crosslinking reaction when exposed to a glow discharge from any of the inert gases. They termed the process "CASING" (Crosslinking by Activated Species of INert Gases). The name was subsequently shown to be a misnomer by Hudis (24) who found that the deep ultraviolet radiation from the discharge causes the reaction. Hence it seems likely that the same mechanism which is responsible for the deep UV stabilization is also active in plasma methods.

The reason for the intrinsic temperature stability of certain resists is probably related to the resist components. Assuming that all positive photoresists contain resin, PAC and solvent as discussed above, we can postulate two ways to increase the stability. The first is to use a solvent which evaporates completely during the spinning or post apply bake. Removing the solvent would raise the viscosity and softening point of the resist. Another method of raising these values is to increase the molecular weight of the resin. As mentioned above, the number average molecular weight of typical resist resins is in the neighborhood of 1000. Gipstein, et al. (25) found that the softening point increased from 95 to 122°C when the molecular weight of the novolak resin increased from 900 to 1350.

Summary

Macroscopic balance equations have been used to investigate the dependence of resist thickness on spinning parameters. The final height of the film is a complex function of initial viscosity and solids content, evaporation rate and residual solvent. The predictions are in qualitative agreement with experiments. As initial viscosity (or solids content) increases, or as the solvent evaporation rate increases, the film thickness increases. In addition the theory indicates that as larger wafers are used, variations in film height may become important because of turbulent mass transfer effects. As long as the fluid is newtonian and the air flow above the wafer is laminar, theory predicts that the thickness is inversely proportional to the square root of the spin speed. This is in good agreement with available data on photoresist spinning. Experimental measurements on polyamic acid solutions show a definite deviation from the inverse square root dependence. Since non-newtonian effects appear to be negligible in the experiments, the cause of this deviation is unknown.

Residual spinning solvent left after the post apply bake, as well as the low molecular weight of the resin used in the resist, lead to deformation of the lithographic images when the resist is heated to temperatures in excess of about 130°C. In order to avoid this problem, a number of image stabilization schemes have been proposed. Both vacuum exposure of the imaged resist to UV light and heating appear to bring about a reaction between the resin and the photoactive compound to produce a high molecular weight material which is insoluble in common solvents and comparatively immobile at high temperatures. If the heating is done slowly, the thermal reaction occurs faster than the image deformation, and some image stability is achieved. Exposing the imaged resist to deep UV radiation or a glow discharge forms a shell around the images which hold them together during the high temperature bake. The mechanisms of these reactions are unknown.

Nomenclature

a, constant in mass transfer expression $(m/s^{1/2})$
C, concentration (kg/m^3)
e, evaporation rate $(kg/cm^2 s)$
h, film height (m)
k, mass transfer coefficient (m/s)
K, distribution coefficient
q, flow rate per unit circumference (m^2/s)
r, radial position (m)
t, time (s)
x, mass fraction
y, dimensionless film height
α, dimensionless evaporation coefficient
$\dot{\gamma}$, shear rate (s^{-1})
ν, kinematic viscosity (m^2/s)
μ, shear viscosity $(Pa_s s)$
ρ, total density (kg/m^3)
τ, dimensionless time
ω, spin speed (rpm)
Ω, spin speed (s^{-1})

Subscripts

f, final value
N, non-volatile component
0, initial value
R, residual value
s, solvent
V, volatile component
∞, ambient atmosphere

Literature Cited

1.   A. G. Emslie, F. T. Bonner, and L. G. Peck, "Flow of a viscous fluid on a rotating disk," J. Appl. Phys. 29 858 (1958)
2.   A. Acrivos, M. J. Shah, and E. E. Petersen, "On the flow of a non-newtonian liquid on a rotating disk," J. Appl. Phys. 31, 963 (1960)
3.   B. D. Washo, "Rheology and modeling of the spin coating process," IBM J. Res. Dev. 21,190 (1977)
4.   D. Meyerhofer, "Characteristics of resist films produced by spinning," J. Appl. Phys. 49, 3993 (1978)

5. F. Krieth, J. H. Taylor, and J. P. Chong, "Heat and mass transfer from a rotating disk," J. Heat Trans. 81, 95 (1959)
6. W. J. Daughten and F. L. Givens, "An investigation of the thickness variation of spun-on thin films commonly associated with the semiconductor industry," J. Electrochem. Soc. 129, 173 (1982)
7. B. T. Chen, "Investigation of the solvent-evaporation effect on spin coating of thin films," Polym. Eng. Sci. 23,83 (1983)
8. F.H. Dill and J.M. Shaw, "Thermal effects on the photoresist AZ1350J," IBM J. Res. Develop. 28, 210 (1977)
9. J.M. Shaw, M.A. Frisch and F.H. Dill, "Thermal analysis of positive photoresist films by mass spectrometry," IBM J. Res. Develop. 28, 219 (1977)
10. J. H. Lai, "An investigation of spin coating of electron resists," Polym. Eng. Sci. 19, 1117 (1979)
11. "Kodak Micro Positive Resist 820," Kodak Publication No. G-103 (1983)
12. S. A. Jenekhe, "The rheology and spin coating of polyimide solutions," Polym. Eng. Sci. 23, 830 (1983)
13. C.G. Wilson, "Organic resist materials-Theory and chemistry," in L.F. Thompson, C.G. Wilson and M.J. Bowden (eds.), Introduction to Microlithography, ACS Symposium Series 219, American Chemical Society, Washington, DC (1983)
14. J. Pacansky and J.R. Lyerla, "Photochemical decomposition mechanisms for AZ-type photoresists," IBM J. Res. Develop. 23, 42 (1979)
15. C.J. Tracy and R. Mattox, "Mask considerations in the plasma etching of aluminum," Solid State Technol. 25:6, 83 (1982)
16. M. Hatzakis and J.M. Shaw, "Diazo-type photoresist systems under electron beam exposure," Electrochemistry Society Meeting, Extended Abstracts, p.927, Seattle, WA (1978)
17. H. Hiraoka and J. Pacansky, "UV hardening of photo- and electron beam resist patterns," J. Vac. Sci. Technol. 19, 1132 (1981)
18. R. Allen, M. Foster and Y.-T. Yen, "Deep UV hardening of positive photoresist patterns," J. Electrochem. Soc. 129, 1379 (1982)
19. H. Hiraoka and J. Pacansky, "High temperature flow resistance of micron sized images in AZ resists," J. Electrochem. Soc. 128, 2645 (1981)
20. F.A. Vollenbroek, E.J. Spiertz and H.J.J. Kroon, "Profile modification of resist patterns in optical lithography," Polym. Engr. Sci. 23, 925 (1983).b 21. J.M. Moran and G.N. Taylor, "Plasma pretreatment to improve resist properties by reduction of resist flow during postbake," J. Vac. Sci. Technol. 19, 1127 (1981)
22. R.H. Hansen and H. Schonhorn, "A new technique for preparing low surface energy polymers for adhesive bonding," J. Polym. Sci b4, 203 (1966)
23. H. Schonhorn and R.H. Hansen, "Surface treatment of polymers for adhesive bonding," J. Appl. Polym. Sci. 11, 1461 (1967)
24. M. Hudis and L.E. Prescott, "Surface crosslinking of polyethylene produced by the ultraviolet radiation from a hydrogen glow discharge," Polym. Letters 10, 179 (1972)
25. E. Gipstein, A.C. Ouano and T. Tompkins, "Evaluation of pure novolak cresol-formaldehyde resins for deep UV lithography," J. Electrochem. Soc. 129, 201 (1982)

RECEIVED December 26, 1984

Effects of Developer Concentration on Linewidth Control in Positive Photoresists

Tom Batchelder

West Coast Technology Center, GCA Corporation, Sunnyvale, CA 94086

As 1μm and submicron processes move from laboratory to pre-
production development, the focus changes from "capability" to
"repeatability" and finally to "controllability and throughput".
It is toward improvements in these last two categories as they
pertain to photoresist process that this paper will address. In
order to restrict the scope of discussion it will be assumed that in
addition to smaller geometries on VLSI chips, that several other
process design constraints will become standard practice for either
technical or economic reasons, discussed below.

The first of these constraints is "in-line process" for all
imaging steps. This leaves considerable latitude for the details of
baking resist coatings and development techniques but includes the
critical limitation that development time must be compatible with
exposure throughput at the camera. As the VLSI era matures the
pressures of automation will both increase the cost of equipment and
demand smooth flow of material from resist coating through development of the image. If processes can be designed to meet this
limitation a vital ancilary benefit will be reduced exposure to
particulate contamination.

The second constraint is that on critical mask levels nearly
monochromatic exposure will be required in order to obtain high
quality images. The requisite reduction in photon flux will lead
to camera throughputs in the neighborhood of 60 wafers per hour.
Tight registration limitations may further limit this number but it
will be used as a reasonable throughput upper limit for VLSI wafers
in next four to five years.

Nearly monochromatic exposure emphasizes a second problem even
as it creates increased image quality, namely: increased standing
waves at the image edge. There are two techniques for minimizing
this problem: multistep baking (2) of resist (before and after image
creation) and the utilization of substrate treatment which renders
it non-reflective(3). It is not clear which of these techniques
will give both high reliability and lowest process cost. The reason
for this is that several other variables, which may differ in a

0097-6156/85/0290-0108$06.00/0

given IC manufacturing process must be considered, namely: step height to be printed over, substrate reflectivity and roughness, minimum feature size to be printed and desired camera throughput. In all but the most severe cases, it is likely that appropriate adjustment of hotplate bake processes before and after imaging will be sufficient to create reliable process down to about 1μm minimum feature sizes. In addition full utilization of process control capabilities of automated in-line wafer development techniques must be made.

Thus, the above constraints limit total development time available from as much as 2 minutes with batch development(1) to development processes of about 30 to 60 seconds. If the γe(6) value is used as a measure of process quality (which is only partially the case as discussed below), it can be seen (Figure 1) that dilute developer gives the best performance. However, the penalty is that required exposure time and thus camera throughput decreases (Figure 2). In addition, since the dose required to obtain breakthrough (Eo) is not only increased but has a steep slope for short development time, linewidth control might be expected to degrade as well. This, infact, has been our experience in practice. In order to obtain equivalent linewidth control with Microposit 351(5) (3.5:1 and 5:1 dilution), the total spray develop time had to be increased from 30 to 60 seconds.

Theory and Experiment

Linewidth over wafer topology ("steps") is affected by three separate physical effects: bulk effect (thickness variation at a step edge), interference effects and undesired reflection from the step or other nearby feature. Only the bulk effect and interference (or standing wave) effect are amenable to clear mathematical treat-ment. The interference effect can be minimized by antireflective films under the resist coating or its results can be minimized by post exposure bake which slightly redistributes the solubility inhibitor at the image edge and/or inhibits development at the image edge due to crosslinking. In any case, reflections from the sub-strate or the step edge cannot be eliminated by development process, developer concentration or photoresist contrast enhancement. These phenomena are optical problems and must be dealt with as part of the exposure process. On the other hand, the bulk effect is directly related to development process and resist contrast. It has been found that(1), for submicron imaging, it is the dominant cause of linewidth variation (0.5μm vs. 0.2μm) over steps on aluminum sub-strates. Thus, while reflection effects must be carefully managed there is still substantial benefit to be gained by optimizing development process. If θ, the edge wall angle is constant for the full resist thickness, it can be shown that(1):

$$(1)\quad \tan\theta = \frac{K\, Zo\, \gamma e}{1 + {}^{\alpha}\, \gamma e\, Zo}$$

where K = imaging constant
Zo = resist thickness
γe = resist contrast
α = resist absorptivity

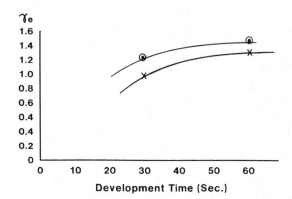

Figure 1. Contrast vs. development time for Microposit 351 developer at 3.5:1 (X) and 5:1 (⊙) concentrations. The Microposit 1470 resist thickness is 10,900 Å, and γ_e is the contrast factor using natural logarithms.

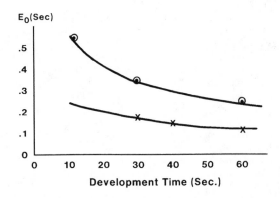

Figure 2. The dose required for breakthrough to the substrate (E_0) vs. development time for Microposit 351 developer at 3.5:1 (X) and 5:1 (O) concentrations. The Microposit 1470 resist thickness is 10,900 Å, and E_0 is the dose required to remove the resist.

and since $\Delta\omega_b$ = 2h/tan θ
  where $\Delta\omega_b$ = change in linewidth due to resist thickness change
        h = height of step
then,
(2) $\Delta\omega_b$ = 2hk $(\dfrac{1}{Z_0 \gamma_e + \alpha})$

Two things are clearly evident from these equations. One is that as the edge wall angle approaches 90°, it becomes increasing resistant to further improvement especially if the absorptivity of the resist is non-zero. Secondly, the absorptivity of the resist limits the possible improvement in linewidth control by means of improvement in γ_e. In the case of the interference effect contribution to linewidth variation, however, a large α value acts to reduce the amplitude of the interference fringes and thus reduces the linewidth variation due to this effect. If an underlying antireflective coating is used in addition to photoresist, the absorptivity, α, of the photoresist should be minimized by the resist manufacturers in order to obtain optimum resist performance. In cases where no antireflective coating is used the value of α at the exposure wavelength may have an optimum value for a given substrate but, at the very least, the absorptivity should be maintained constant at the exposure wavelength by resist manufacturers.

In a recent paper W.D. Hinsberg(4) demonstrated that the Na+ ion concentration plays a key role in the dissolution rate of both exposed and unexposed photoresist. While the exact machanism of this effect is not yet known, it was possible to obtain a function dissolution rate equation rate for unexposed resist(4):

(1) $Ru = 2.3 \times 10^5 [Na+] [OH-]^{3.7}$
   where
   Ru is the dissolution rate for unexposed resist
   [Na+] is the molar concentration of sodium ion
   [OH-] is the molar concentration of hydroxide ion

If the [OH-] is restricted to that normally obtained with commercially available AZ351 developer at various normally used dilutions (~.05-.25 molar in OH- concentration), it is possible to make a linear approximation to the data obtained by Hinsberg for the conditions of exposure defined by Hinsberg and a dissolution rate equation for exposed resist:

(2) $Re \approx 2.4 \times 10^4 [Na+] [OH-]^{.6}$
   where
   Re = the dissolution rate for positive resist after a 90MJ/cm^2
        exposure dose at measured at a point 1μm below the surface
        of a 2μm thick resist film.
The ratio:
(3) $Re/Ru \sim .1/[OH-]^3$

From this data it is possible to show, for hydroxide ion and sodium ion concentrations encountered in typical dilutions fo the Microposit series of developers (3.5 to 1 and 5 to 1 dilutions of undiluted developer).

The solubility rate ratio is a commonly used value to measure

relative contrast of resist for a given film with a given dose. The
key thing to note is that this ratio does not show any [Na+] dependance. Thus the γe value should be roughly independent of [Na+].
γe values for Microposit 351 at 3.5 to 1 and 5:1 concentration were
measured and found to be nearly independent of [Na+]. Similarly the
linewidth control for microposit developers was found to be independent of sodium ion concentration.

     Now, since the exposed area's dissolution rate is dependent on
sodium ion concentration, (as seen in "Equation 2") this suggests a
straight forward method for optimizing both the contrast (and thus
the linewidth control) and the development rate. Optimum performance (high γe value and linewidth control) can be obtained by
using dilute developers and the absolute dissolution rate loss can
be re-established by means of adjusting the sodium salt (e.g. NaCl)
concentration. This proved to be the case.

     The images obtained with Microposit 351 (5:1) dilution with no
added NaCl, 0.5 moles NaCl, and 1.0 moles of NaCl are shown in
Figures 3, 4, 5. It can be seen that the required relative exposure
time dropped substantially from 640 msec. to 390 msec. When 0.5
moles of NaCl was added to the developer, the linewidth control
across the wafer was unchanged. The prediction from equation(3) can
be seen to be qualitatively confirmed. A similar effect was observed
with Microposit 351 at the 3.5 to 1 concentration. The reduction in
photospeed at higher concentration of NaCl remains unexplained. However, it suggests that the key to optimum developer speed is Na+ ion
concentration and optimum contrast is obtained primarily by adjustment
of the OH$^-$ ion concentration.

     In order to prove that sodium chloride by itself showed no selec-
tivity for exposed or unexposed resist, an exposed wafer was soaked in
a 2 molar solution of sodium chloride for 10 minutes. No resist loss
was noted in either exposed or unexposed resist. In addition, if the
sodium ion is in sufficient concentration it no longer enhances the
reaction but serves to mask the hydroxide attack on the carboxylic
acid sites.

     In order to confirm the effect predicted from the equation ob-
tained by utilizing of data on sodium hydroxide/sodium chloride solutions, experiments were performed with the buffered Microposit series
of developers (Na OH solution with Na_3BO_3 buffer) to obtain the Re
(dissolution rate of exposed resist) and Ru (dissolution rate of unexposed resist). The results of these measurements are shown in
Figures 6, 7, 8 as function of sodium chloride concentration. It can
be seen that the effect of the ½ molar sodium chloride solution is to
enhance the dissolution rate of both the exposed and unexposed photoresist areas. When the ratio of Re/Ru is obtained (Table I), it is
clear that the 5:1 developer dilution has significantly higher contrast and that the ratios are roughly constant for all sodium chloride
concentration studied.

     In the course of these studies, it was noted that the primary
effect on mean dissolution rate was near the wafer surface. Namely, on
those developers in which NaCl was added, the dissolution rate for
short develop time attained the equilibrium value much more quickly
than with those in which no NaCl was added (See Figure 7 and 8) . In
order to confirm this observation the development initiation time was
measured interferometrically. Development initiation time

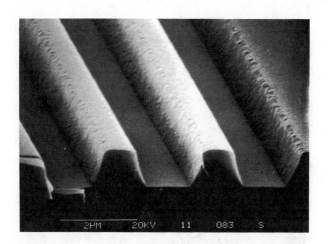

Figure 3. Image obtained with Microposit 351 developer (5:1   dilution) with no added NaCl. Control is resist 1400.27; the exposure time is 0.640 ms; the development time is 30 s; and softbake is at 75 °C/45s.

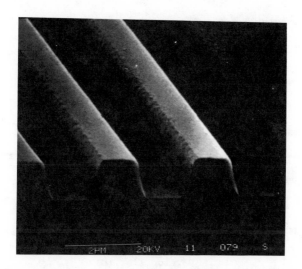

Figure 4. Image obtained with Microposit 351 developer (5:1 dilution) with 0.5 mol of NaCl added. The exposure time is 0.390 ms; the development time is 30 s; softbake is at 75 °C/45 s; and postexposure bake is at 110 °C/45 s.

Figure 5. Image obtained with Microposit 351 developer (5:1 dilution) with 1.0 mol of NaCl added. The exposure time is 0.630 ms; the development time is 30 s; softbake is at 75 ^{o}C/45 s; and postexposure bake is at 110 ^{o}C/45 s.

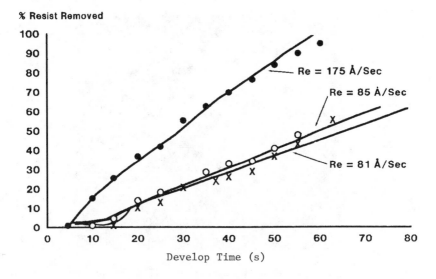

Figure 6. Percent resist removed vs. development time for Microposit 351 developer (5:1 dilution) with 0 (X), 0.5 (●), and 1.0 (o) mol/L of NaCl added to the developer. Exposure time is 180 ms.

% Resist Removed

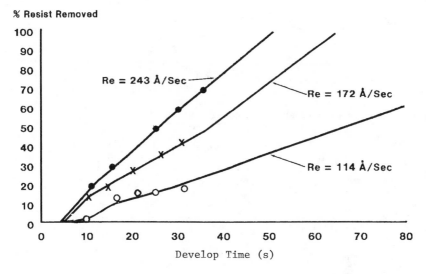

Figure 7. Percent resist removed vs. development time for Microposit 351 developer (3.5:1 dilution) with 0 (X), 0.5 (●), and 1.0 (o) mol/L of NaCl added to the developer. Exposure time is 120 ms.

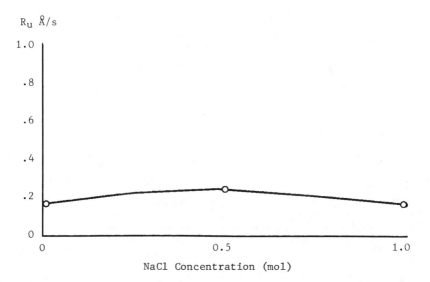

Figure 8. Dissolution rate vs. NaCl concentration for an unexposed resist with a Microposit 351 developer concentration of 5:1 and 10 min in the developer.

Table I. Dissolution Rate Ratios for Microposit 351
Developer at 3.5:1 and 5:1 Dilution

Concentration	NaCl Concentration	R_e	R_u	R_e/R_u
3.5:1	0.0 Moles	172	2.5	69
	0.5	243	2.2	110
	1.0	114	1.2	95
5:1	0.0 Moles	81	.16	506
	0.5	175	.31	565
	1.0	85	.15	566

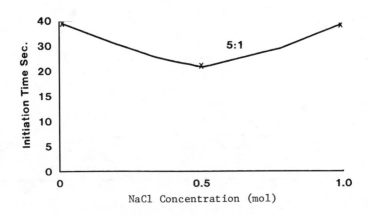

Figure 9. Initiation time vs. NaCl concentration in Microposit 351
developer (5:1 dilution).

was defined as the time after the exposed wafer was placed in the developer before any measurable photoresist loss could be detected. This effect would of course be most pronounced in underexposed resist and dilute developers. The results are seen in Figure 9. Clearly, the pronounced reduction in initiation time is a primary cause of increased mean dissolution rate or conversely, the reason why required exposure time is greatly reduced.

Summary

It has been shown that both OH⁻ and Na+ ion concentration play a key role optimizing a developer performance. The primary effect of Na+ ion is to enhance carboxylic acid/hydroxide dissolution mechanism. This effect is most pronounced at the resist surface where significant reduction in development initiation time was noted. If additional Na+ ion is added, it finally only serves to mask the OH⁻ ion attack on the carboxylic acid sites. Thus photoresist contrast (γe) may be optimized by developer concentration and developer speed may be optimized independently by addition of NaCl or other Na+ ion containing salt.

Literature Cited

1. Arden, W.; Keller, H.; Mader, L.   Solid State Technology, July 1983, pp. 143–150.
2. Batchelder, T.; Piatt, J.   Solid State Technology, August 1983 pp. 211–217.
3. Coyne, R.; Brewer, T.   Kodak Microelectronics Seminar Interface 1983, November 1983.
4. Hinsberg, W.D.; Gutierrez, M.L.   Kodak Microelectronics Seminar Interface 1983, November 1983.
5. Microposit™351 developer is product of Shipley Co. and is used commonly as a developer for Microposit™ 1400 series resists. It consists of a Na OH solution with Na_3BO_3 buffer.
6. The contrast factor, γe, is defined as the slope of the natural logorithm vs. % resist thickness remaining curve measured at the resist/substrate interface after the convention of W. Arden et.al., in Reference 1 above.

RECEIVED March 29, 1985

The Advantage of Molecular-Beam Epitaxy for Device Applications

A. Y. Cho

AT&T Bell Laboratories, Murray Hill, NJ 07974

Continuous striving to improve the performance of semiconductor devices motivated much research in semiconductor material science. In turn, a new material may generate new devices and new technology. Molecular beam epitaxy may hold the key to producing the next generation microwave and optoelectronic devices with "band-gap engineering." Monolithic integration and in situ dry processing will be unique developments of MBE for the coming years.

The most important feature of molecular beam epitaxy (MBE) is its flexibility in growing various compounds and structures with precise doping and compositional profiles accurate to atomic dimensions. This is one of the reasons that MBE has led in the technology forefront such as with the first fabrication of multilayered structures with dimensions smaller than a carrier diffusion length (1,2). Unique electrical (3,4) and optical (5) properties of MBE grown layers that do not exist in bulk semiconductors have led to developments such as quantum well lasers, (6,7) modulation doped high electron mobility devices, (8−10) and separately controlled absorption and multiplication region (SCAM) photodiodes (11). In this communication, we would like to discuss the advantages of MBE for device fabrication, the present status, and future developments.

Advantages of MBE

Molecular beam epitaxy is conducted in an ultra high vacuum environment (12,13). The film material (constituent) can therefore be started or stopped abruptly to grow thin layers with abrupt changes in doping and compositional profiles. Most of the devices made to date resulted from the fact that semiconductor materials were grown in one reactor and the metallization was done in another. Between these two steps, more than 30Å of oxide is usually formed on the surface of the semiconductor. Since MBE is a vacuum process, many of these device fabrication processes can be incorporated with the MBE system as shown in Fig. 1. For instance, for the purpose of making smaller devices with higher yields, non-alloyed ohmic contacts, (14,15) Schottky barrier diodes, (16,17) insulating layers, (18−20) and in situ masking (21−24) can all be performed in the MBE system. A combination of MBE and ion implantation may produce precise dimensions in both lateral and vertical directions (25). The epitaxial growth rate with MBE can be varied from 0.1 μm/hr to 10 μm/hr. The advantage of a slower growth rate compared to that of chemical vapor deposition or liquid phase epitaxy is that one can control the layer thickness more precisely. The fast production rate of MBE depends upon its fast

0097-6156/85/0290-0118$06.00/0

turnaround time. For example, one may load six to twelve wafers with 2 inch diameter in a commercial MBE system, and the time required for exchanging substrates is less than five minutes. No bakeout or gas purging is required for MBE between sample exchanges. One may therefore prepare twelve field effect transistor (FET) wafers or six laser wafers in a day. This one-day production may keep a twenty person processing line occupied for one week.

Molecular beam epitaxy is very versatile. One may grow an FET, a laser, a mixer diode, a bipolar transistor, a photodetector, and a varactor in one sequence and in one day without modifying either the sample holder or the deposition geometry. The epitaxial growth is not under thermodynamic equilibrium and therefore a large range of abrupt changes in doping profile and compositional profile are possible. Furthermore, nucleation on a foreign substrate is more readily accomplished than with other growth techniques. For instance, the growth for a lattice mismatched system will proceed even when the strain energy exceeds the Gibbs energy. The controlled strained superlattice is another new development with MBE.

One other important point is the conservation of film materials. For the growth of GaAs, ten grams of Ga will prepare more than 100 FET wafers. A point that becomes of increasing concern is operational safety. MBE may be operated with solid sources (Ga, As, Al, P, Si, Sn, Be, ...) with refill load-lock for ovens without using toxic gases such as arsine or phosphine. However, there is some advantage in using the latter for $Ga_xIn_{1-x}As_{1-y}P_y$ which will be discussed in a following section.

Present Development of MBE

One of the recent efforts in MBE for device fabrication has been to extend the growth to materials such as $In_xGa_{1-x}As$, (26–28) $In_xAl_{1-x}As$, (29,30) $Ga_xAl_yIn_{1-x-y}As$, (31,32) and $Ga_xIn_{1-x}As_{1-y}P_y$ (33–36). These alloy systems cover the band-gap energies (wavelengths) from 0.756 eV (1.65 μm) to 1.55 eV (0.8 μm) and are lattice matched to InP substrate at the same time. For the $Ga_xAl_yIn_{1-x}As$ system solid charges of Ga, Al, In, and As may be used (18) while the $Ga_xIn_{1-x}As_{1-y}P_y$ system "gas source" MBE appears to be the best choice (35,37-38).

One of the most intriguing manners by which to achieve a laser operating at 1.55 μm with $Ga_{0.47}In_{0.53}As$ is to fabricate quantum-well lasers instead of the standard double-heterostructure lasers which would operate at 1.65 μm. The first successful preparation of a $Ga_{0.47}In_{0.53}As/Al_{0.48}In_{0.52}As$ multiquantum-well laser was recently reported (39). The schematic diagram of the quantum-well laser structure and the spontaneous and stimulated emission characteristics are shown in Fig. 2. The active layer consisted of a series of $Ga_{0.47}In_{0.53}As$ wells separated by barriers of $Al_{0.48}In_{0.52}As$. The large band-gap difference between wells and barriers, $\Delta Eq = 0.7$. eV and the large conduction band discontinuity, $\Delta E_c = 0.5$ eV, (40) result in very efficient carrier confinement. For a sample with 13 layers of $Ga_{0.47}In_{0.53}As$, 90Å thick, separated by 12 layers of $Al_{0.48}In_{0.52}As$, 30Å, one may estimate the up shift in lasing energy to be 56 meV. Adding the heavy hole contribution of ≈10% of the above, the energy shift is expected to be about 63 meV. The emission peak in Fig. 2 is in agreement with the calculated values. Figure 3 shows the light output as a function of current for this multiquantum-well laser. Threshold current density as low as 2.4 KA/cm^2 was achieved (39).

Excellent results of double heterostructure and separate confinement heterostructure lasers of $Ga_xIn_{1-x}As_{1-y}P_y$ lattice matched to InP and emitting at 1.5 μm have been reported recently (36). These lasers were grown by gas source MBE utilizing the decomposition of AsH_3 and PH_3 at 900-1000°C as a source of As_2 and P_2 molecules. The advantage of using gas sources is to reduce the frequency for group V element oven replenishment if solid group V sources were used. It is also more reproducible in achieving precise As_2/P_2 ratios by the use of precise gas flow controls. Source tanks containing 250 gm of 100% PH_3 and 50 gm of 100%

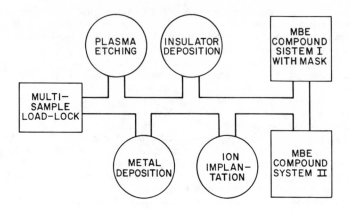

Figure 1.   Conceptual arrangement of an in situ processing MBE system.

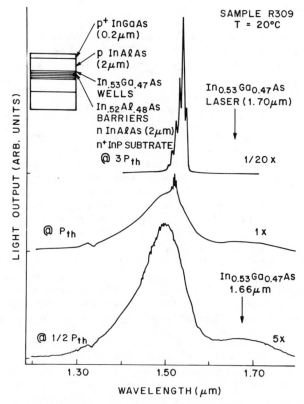

Figure 2.   Spontaneous and stimulated emission of a quantum–well laser with 80-90 Å wells obtained by optical pumping.   (Adapted from Ref. 39.)

AsH$_3$ can prepare approximately 360 laser wafers. A schematic of the gas source cracking oven and tne gas handling system is shown in Fig. 4. At a growth temperature of about 525°C, the accommodation coefficient of P is less than that of As on the growing surface (35,41,42). Therefore, a higher P to As ratio is required in the beam flux than that in the solid. This relationship is illustrated in Fig. 5 (Ref. 35). Broad area lasers ($\lambda = 1.5$ μm) with room temperature pulsed threshold current densities of 2 KA/cm^2 were reproducibly obtained (36).

Another new development in MBE is to demonstrate the capability for simultaneously lattice matched graded composition structures in the quaternary material system (32). This "band-gap engineering" resulting from the graded composition opens a new window in our search for a new degree of freedom for device and structure design. Grading of the high field region and superlattices in an avalanche photodiode have been used to enhance the ionization rates ratio in the GaAs/AlGaAs system (43). Recently, the first vertical Npn Al$_{0.48}$In$_{0.52}$As/Ga$_{0.47}$In$_{0.53}$As heterojunction bipolar transistor with a graded emitter comprised of a quaternary layer of Ga$_{0.47-x}$Al$_x$In$_{0.53}$As was reported (32). Grading from Ga$_{0.47}$In$_{0.53}$As to Al$_{0.48}$In$_{0.52}$As was achieved by simultaneously lowering the Al and raising the Ga oven temperature in such a manner as to keep the total group III flux constant during the transition. The energy band diagram for the abrupt and graded transistors are shown in Fig. 6(a) and 6(b), respectively. The elimination of the spike at the heterojunction for the graded emitter case enhances the forward injection and results in an improvement of current gain from 140 to 280 as illustrated in Fig. 7(a) and 7(b), respectively.

Molecular beam epitaxy has the ability to grow extremely sharp doping profiles with n+ and p+ layers. One therefore can design a separately controlled absorption and multiplication region (SCAM) photodiode (11). The electric field profile in the avalanche region is controlled by the hi-lo doping profile resembling a hi-lo IMPATT diode (44). The electric field at the heterojunction region for this structure is constant rather than linearly graded. This eliminates the problem of tunneling in the low gap layer. Together with the lower electric field in the gain region it becomes easier to achieve simultaneously both high gain and low dark current. For a typical structure, the high band-gap region is comprised of a 5000Å undoped (10^{13}/cm^3) Al$_{0.48}$In$_{0.52}$As avalanche layer, and a 500Å thick n+ spike doped to 5×10^{17}/cm^3, separated from the heterojunction interface by a low electric field Al$_{0.48}$In$_{0.52}$As i-layer, 2000Å thick. The absorption layer is 2 μm of undoped Ga$_{0.47}$In$_{0.53}$As (n<10^{16}/cm^3). The device was optimized for low voltage operation. The breakdown voltage was 25-27V which is the lowest reported in III-V long wavelength APD's. The avalanche gain was 50 with a dark current of 50 nA for a device 2×10^{-4}cm^2 in area. The external quantum efficiency was 60% at 1.55 μm. See Figure 8.

Future Development of MBE

Multiple processing chambers including plasma etching, electron gun evaporation for in situ metallization, SiN$_4$ deposition or fluoride epitaxial growth for passivation or MIS structures will be incorporated into MBE systems. Multiple level masking in situ of the MBE growth will be developed. For large scale production of Ga$_x$In$_{1-x}$As$_{1-y}$P$_y$ quaternary devices, "gas source" MBE will be incorporated. All the growth processes will be computer automated with 3 to 4 inch substrates. The very long wavelength material (3-12 μm) such as Hg$_x$Cd$_{1-x}$Te will become increasingly important. MBE already has an excellent start in this area (45). Microwave and optoelectronic device integration and in situ dry processing will be the unique developments of MBE for the coming year.

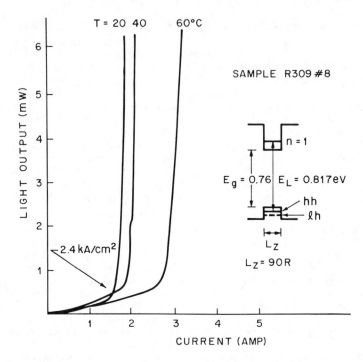

Figure 3. Light output as a function of current of the injection laser operating at 1.51 μm (at 20 °C). Insert shows quantum-well parameters. (Adapted from Ref. 39.)

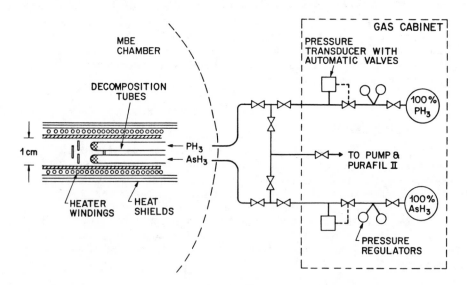

Figure 4. Schematic of a high pressure gas source and the gas handling system for MBE growth. (Adapted from Ref. 35.)

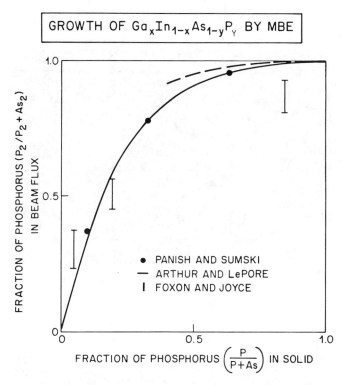

GROWTH OF $Ga_xIn_{1-x}As_{1-y}P_Y$ BY MBE

- PANISH AND SUMSKI
— ARTHUR AND LePORE
I FOXON AND JOYCE

FRACTION OF PHOSPHORUS $\left(\dfrac{P}{P+As}\right)$ IN SOLID

Figure 5. Relative concentration of phosphorus in the molecular beam as a function of that incorporated in the solid. (Adapted from Ref. 35.)

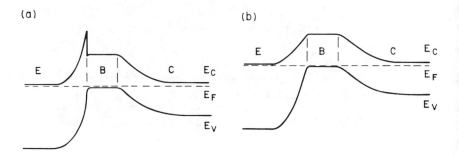

(a)   (b)

Figure 6. Schematic energy–band diagram under equilibrium for the $Al_{0.48}In_{0.52}As/Ga_{0.47}In_{0.53}As$ heterojunction bipolar transistor with (a) abrupt emitter and (b) graded emitter. Note the elimination of the conduction band notch through the use of a graded emitter. (Adapted from Ref. 32.)

(a)

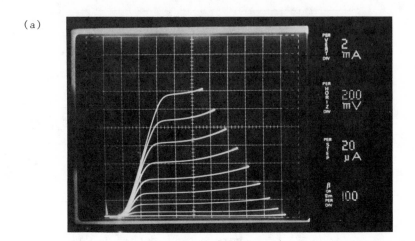

(b)

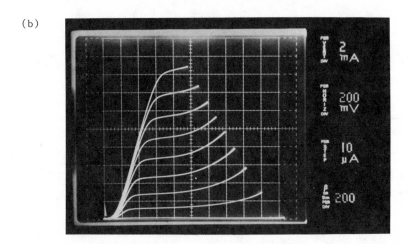

Figure 7. Common-emitter characteristics of the $Al_{0.48}In_{0.52}As/Ga_{0.47}$-$In_{0.53}As$ heterojunction bipolar transistor with (a) abrupt emitter and (b) graded emitter at 300 K. (Adapted from Ref. 32.)

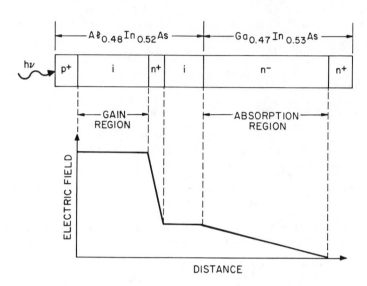

Figure 8.   Schematic of a separately controlled absorption and
multiplication regions (SCAM) avalanche photodiode (APD) using high-
low doping profiles in the avalanche region to produce constant
electric field. (Adapted from Ref. 11.)

Acknowledgments

      The author would like to thank M. B. Panish, F. Capasso, and K. Alavi for useful advice
and discussions.

Literature Cited

1. A. Y. Cho, Appl. Phys. Lett. $\underline{19}$, 467 (1971).
2. L. L. Chang, L. Esaki, W. E. Howard, and R. Ludeke, J. Vac. Sci. Technol. $\underline{10}$, 655
   (1973).
3. L. Esaki and L. L. Chang, Phys. Rev. Lett. $\underline{33}$, 495 (1974).
4. H. L. Stormer, D. C. Tsui, A. C. Gossard, and J. C. Hwang, Physica $\underline{117B}$, 688 (1983).
5. R. Dingle, W. Wiegmann, and C. H. Henry, Phys. Rev. Lett $\underline{33}$, 827 (1974).
6. J. P. van der Ziel, R. Dingle, R. C. Miller, W. Wiegmann, and W. A. Nordland, J. Appl.
   Phys. Lett. $\underline{26}$, 463 (1975).
7. W. T. Tsang, Appl. Phys. Lett. $\underline{39}$, 786 (1981).
8. R. Dingle, H. L. Stormer, A. C. Gossard, and W. Wiegmann, Appl. Phys. Lett. $\underline{33}$, 554
   (1978).
9. S. Hiyamizu, T. Mimura and T. Ishikawa, Jpn. J. Appl. Phys. $\underline{20}$, L-598 (1981).
10. P. N. Tung, D. Delagebeaudeuf, M. Laviron, P. Delescluse, J. Chaplart, and N. T. Linh,
    Elect. Lett. $\underline{18}$, 3 (1982).

11. F. Capasso, K. Alavi, A. Y. Cho, P. W. Foy and C. G. Bethea, International Electron Device Meeting Digest, p. 468, Washington, DC, December 5-7, 1983.
12. A. Y. Cho and J. R. Arthur, Progress in Solid State Chemistry, edited by G. Somorjai and J. 0. McCaldin (Pergammon, New York, 1975), Vol. 10, p. 157.
13. A. Y. Cho, Thin Solid Film 100, 291 (1983).
14. P. A. Barnes and A. Y. Cho, Appl. Phys. Lett. 33, 651 (1978).
15. J. V. DiLorenzo, W. C. Niehaus and A. Y. Cho, J. Appl. Phys. 50, 951 (1979).
16. A. Y. Cho and P. D. Dernier, J. Appl. Phys. 49, 3328 (1978).
17. A. Y. Cho, E. Kollberg, H. Zirath, W. W. Snell, and M. V. Schneider, Elec. Lett. 18, 424 (1982).
18. H. C. Casey, Jr., A. Y. Cho, D. V. Lang, and E. H. Nicollian, J. Vac. Sci. Technol. 15, 1408 (1978).
19. R. F. C. Farrow, P. W. Sullivan, G. M. Williams, G. R. Jones, and D. C. Cameron, J. Vac. Sci. Technol. 19, 415 (1981).
20. C. W. Tu, S. R. Forrest, and W. D. Johnston, Jr., Appl. Phys. Lett. 43 569 (1983).
21. A. Y. Cho and F. K. Reinhart, Appl. Phys. Lett. 21, 355 (1972).
22. W. T. Tsang and A. Y. Cho, Appl. Phys. Lett. 32, 491 (1978).
23. A. Y. Cho and W. C. Ballamy, J. Appl. Phys. 46, 783 (1975).
24. W. T. Tsang and A. Y. Cho, Appl. Phys. Lett. 30 293 (1977).
25. K. Tabatabaie-Alavi, A. N. M. M. Choudhury, K. Alavi, J. Vlcek, N. Slater, C. G. Fonstad, and A. Y. Cho, IEEE Electron. Dev. Lett. EDL−3, 379 (1982).
26. K. Y. Cheng and A. Y. Cho, J. Appl. Phys. 53, 4411 (1982).
27. Y. Kawamura, Y. Moguchi, H. Asaki, and H. Nagai, Electron. Lett. 18, 91 (1982).
28. J. Massies, J. Rochette, P. Delescluse, P. Eitenne, J. Chevrier and N. T. Linh, Electron. Lett. 18, 758 (1982).
29. K. Y. Cheng and A. Y. Cho, J. Appl. Phys. 53, 4411 (1982).
30. M. A. Brummell, R. J. Nicholas, J. C. Portal, K. Y. Cheng, and A. Y. Cho, J. Phys. C: Solid State Phys. 16, L579 (1983).
31. K. Alavi, H. Temkin, W. R. Wagner and A. Y. Cho, Appl. Phys. Lett. 42, 254 (1983).
32. R. J. Malik, J. R. Hayes, F. Capasso, K. Alavi, and A. Y. Cho, IEEE Electron. Dev. Lett. EDL−4, 383 (1983).
33. A. Y. Cho, J. Vac. Sci. Technol. 16, 275 (1979).
34. W. T. Tsang, F. K. Reinhart and J. A. Ditzenberger, Appl. Phys. Lett. 41, 1094 (1982).
35. M. B. Panish and S. Sumski, J. Appl. Phys. (to be published).
36. M. B. Panish and S. Sumski, Appl. Phys. Lett. (to be published).
37. A. R. Calawa, Appl. Phys. Lett. 38, 701 (1981).
38. M. B. Panish, J. Electrochem. Soc. 127, 2729 (1980).
39. H. Temkin, K. Alavi, W. R. Wagner, T. P. Pearsall, and A. Y. Cho, Appl. Phys. Lett. 42, 845 (1983).
40. R. People, K. W. Wecht, K. Alavi, and A. Y. Cho, Appl. Phys. Lett. 43, 118 (1983).
41. J. R. Arthur and J. J. LePore, J. Vac. Sci. Technol. 6, 545 (1969).
42. C. T. Foxon, B. A. Joyce and M. T. Morris, J. Cryst. Growth 49, 132 (1980).
43. F. Capasso, J. Vac. Sci. Technol. B1, 457 (1983).
44. A. Y. Cho, C. N. Dunn, R. L. Kuvas, and W. E. Schroeder, Appl. Phys. Lett. 25, 224 (1974).
45. J. P. Faurie and A. Million, J. Cryst. Growth 54, 582 (1982).

RECEIVED March 12, 1985

Chemical and Physical Processing of Ion-Implanted Integrated Circuits

Joseph C. Plunkett

Department of Electrical Engineering, California State University, Fresno, CA 93740

A brief overview of the fundamentals of the chemical and
physical processing of ion-implanted integrated circuits
is presented. Although not intended as a thorough review paper in the field, a modest list of references are
provided to which the reader may refer for more in-depth
discussions of the topics covered. As well as an overview of the principles of ion-implantation, profile
shaping as a means of improving device performance is
discussed. Typical applications of ion-implantation in
silicon and gallium arsenide devices are also covered.
Finally, some basic clean-up processes for laboratory
ion-implantation processing are provided.

Technologies for the chemical and physical processing of integrated
circuits have advanced at a very rapid pace in recent years. The
stringent requirements of improved electrical performance and further microminiaturization of integrated circuits have forced advancement of the state-of-the-art in fabrication technology. With
the evolvement of sub-micron technology and the extension of monolithic integrated circuit technology to the microwave LSI range,
ion implantation is playing an increasingly important role in the
fabrication processes of solid state circuits and devices. Ion
implantation has been the subject of numerous review articles and
bibliographies in recent years ([1-20]).
 The discussions in this paper provide fundamental and overview
material suitable for the new researcher in the field of integrated
circuit processing, as well as provides the user of ion implantation
with some relevant design information and some recent applications
to some new devices and materials.
 In its most fundamental form, ion implantation is a process by
which energetic, charged-particles or impurity atoms can be introduced into a target or substrate material. Often, these particles
which are to be implanted are positive ions (singly or multiply
ionized) which come from a suitable source. After the formation of
the positive ions, they are accelerated by static electric fields,
are focused or formed into a beam, and are passed through a mass
analyzer. The mass analyzer is used to ensure that the beam is pure.

0097-6156/85/0290-0127$10.00/0
© 1985 American Chemical Society

The beam is often further manipulated by slits or quadrupole ana-
lyzers prior to its striking the target. As applied to the doping
of materials, the ion beam must usually be made to form a raster
scan of the target. The ion beam then bombards the target substrate
material by a carefully controlled and quantified process whereupon
most of the ions enter the substrate material. Upon the introduction
of the ions into the target material, the material acquires new electrical or chemical properties.

A brief overview of some of the advantages and disadvantages of
ion implantation processes in the doping of solid state semiconductor
materials is presented below.

(1) Variety of sources--for most semiconductor applications the
ions which are introduced must be electrically active. This means
that for impurity doping of a substrate material the ions must occupy
substitutional sites in the lattice structure.

(2) Variety of substrate material--although virtually any material can be implanted, one must choose a material for impurity doping
in solid state semiconductors which can be electrically activated.

(3) Impurity concentration profiles can be shaped to certain
specifications. One important aspect of ion implantation into
semiconductors, in contrast to the diffusion process, is that the
number of implanted atoms can be precisely controlled by the external system, rather than by the physical parameters of the target
material.

(4) Ordinary photolithography or mechanical masking is used for
impurity positioning in the substrate.

(5) Ion implantation is a low temperature process which will not
usually disturb earlier impurity distributions which may have been
placed in the material. An anneal cycle must, however, be considered
in order to electrically activate the dopant atoms.

(6) Implantation is not solubility limited--often however, during the process of electrical activation a condition of equilibrium
may be reached and precipitation of the excess impurities may occur.

(7) Electrical activation can be achieved at temperatures less
than that required for diffusion in certain materials, although
gallium arsenide requires a rather high temperature. In silicon,
a nominal temperature of 600 to 800 degrees C is used.

(8) The process of ion implantation lends itself to automatic
control. With the advent of microprocessors and microcomputers,
such process control as profile shaping can be programmed into the
control and made a routine and precise process.

Ion Distribution and Penetration

Ion implantation is not an equilibrium process, although in the
process of electrical activation, equilibrium may be achieved. The
energetic ions lose their energy to the host lattice often creating damage to the bulk material. An annealing process is used to
remove most of the damage and to electrically activate the ions.
The loss of energy to the substrate material is brought about by two
processes. One is by excitation and ionization of electrons. The
other is by elastic collisions with nuclei. The results of violent
collisions may be the displacement of atoms in the host lattice along
the ion path setting up a chain reaction as long as the kinetic

energy is available. Although some of the damage to the material
sets up an amorphous state, annealing can return the lattice to its
crystalline form.
     Certain critical alignments of the ion beam can cause channel-
ing or guiding of the ions by the crystal lattice. This can cause
deep anomolous penetration of the substrate if they are critically
aligned along the axes of the target crystal.
     In practice, the target is usually deliberately misaligned with
respect to the major axes of the crystal. This misalignment angle
is typically 7 to 10 degrees off-axis. This quenches the channeling
effect and simulates an amorphous target.

Theoretical Range Determination. The LSS theory (21) is often used
to calculate the theoretical range, $\bar{R}$, and total straggle ΔR. The
straggle represents the statistical fluctuation of the total range.
It is assumed that an amorphous target is used and that a Gaussian
distribution is created. Figure 1 illustrates the depth distribution of implanted atoms in an amorphous target. In the first curve,
the incident ion mass is less than the substrate atomic mass. In
the second curve, the ion mass is greater than the host material
atomic mass. As can be seen, the straggle is greater for the first
case.
     For practical solid state device doping, the mean perpendicular
depth of penetration, $\bar{R}_p$, and the associated straggle, $\Delta \bar{R}_p$, are the
important parameters. For critical masking control, however, trans-
verse straggle, ΔR_t, can be important.
     Figure 2 shows the reference coordinates and nomenclature of
this geometry. Theoretical calculations of the projected range and
straggle for various dopants and substrates have been calculated and
shown in graphical form (4, 21-26). Some common dopant ions for
silicon and gallium arsenide are shown in Figures 3, 4, 5, and 6.

Theoretical Impurity Profile Calculations. Using information provided by the preceding curves, a theoretical as-implanted concentration profile can be calculated in terms of the preceding parameters (from Stone and Plunkett (26) and references therein).

$$n(x,y) = \frac{s}{(2\pi)^{3/2} \Delta \bar{R}_p \Delta \bar{R}_t^{2}} \exp\left[-\left[\frac{x - \bar{R}_p}{(2)^{1/2} \Delta \bar{R}_p}\right]^2\right]$$

$$x \exp\left[-\left[\frac{y}{(2)^{1/2} \Delta \bar{R}_t}\right]^2\right] \tag{1}$$

where the coordinates are shown in Figure 2.
n is the ion concentration for s ions/unit surface perpendicular
to the target surface in the x direction.
     For typical applications, the ΔR_t parameter can be effectively
eliminated by the beam scan which creates an overlap of the implanted
ions over the transverse straggling range. Of course, the lateral
straggling effect cannot be eliminated at the mask edges.

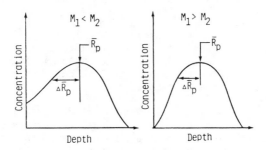

Figure 1. The depth distribution of implanted atoms in an amorphous target.

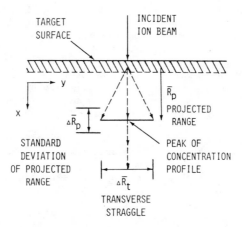

Figure 2. Reference coordinates for ion implantation parameters.

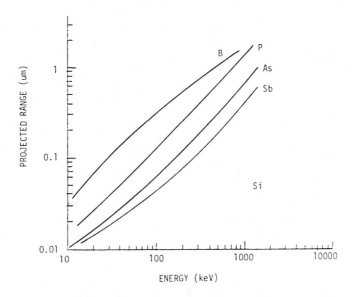

Figure 3.   Theoretical calculations of the projected range of
B, P, As, and Sb in silicon. (Adapted with permission from
Reference 27, copyright 1983, John Wiley and Sons).

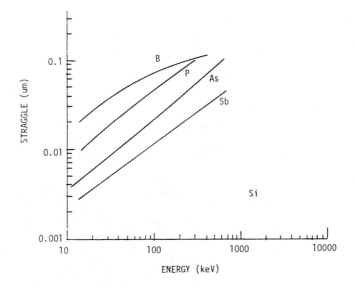

Figure 4.   Theoretical calculations of projected straggle for
B, P, As, and Sb in silicon. (Adapted with permission from
Reference 27, copyright 1983, John Wiley and Sons).

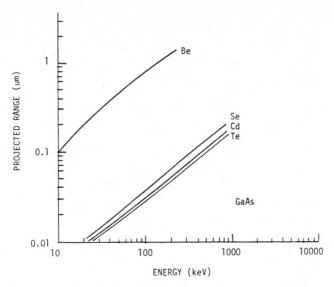

Figure 5.   Theoretical calculations of the projected range of
Be, Se, Cd, and Te in GaAs.   (Adapted with permission from
Reference 27, copyright 1983, John Wiley and Sons).

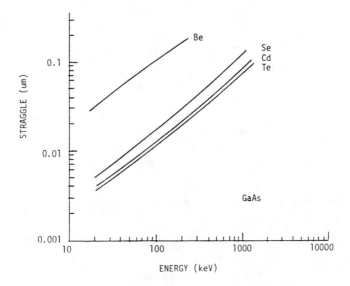

Figure 6.   Theoretical calculations of the projected straggle of
Be, Se, Cd, and Te in GaAs.   (Adapted with permission from
Reference 27, copyright 1983, John Wiley and Sons).

In the primary area of the scan, the profile expression thus becomes:

$$n(x) = N_{max} \exp\left[-\frac{(x - \bar{R}_p)^2}{2\,\Delta\bar{R}_p{}^2} \right] \qquad (2)$$

where N_{max} is the peak of the concentration profile occurring at the point $x = \bar{R}_p$:

$$N_{max} = \frac{N_\square}{(2\pi)^{1/2}\,\Delta\bar{R}_p} \approx \frac{0.4\,N_\square}{\Delta\bar{R}_p} \qquad (3)$$

$N_\square$ is defined as the ion dose in ions/cm^2. It is related to the total implanted charge Q in C/cm^2 by

$$N_\square = Q_\square/q \qquad (4)$$

where q = charge on the electron.

The total implanted charge, Q, can be experimantally determined by integrating the beam current impinging on the target during the implantation time. If the beam current, I, is constant, and for implant time $\underline{t}$:

$$Q = \frac{(I)(t)}{A} \qquad (5)$$

where A is the target area scanned by the ion beam. This quantity when combined with the known energy of the accelerated ions, the theoretical values for $\bar{R}_p$ and ΔR_p can be determined from the various curves presented. This allows the theoretical impurity concentration curves to be constructed.

An example is used to illustrate the application of the above theory. Assume that a five inch diameter silicon wafer is uniformly implanted with 80 keV boron atoms for five minutes with a constant beam current of ten microamperes. Then

$$N_\square = \frac{Q_\square}{q} = \frac{(I)(t)}{qA} \qquad (6)$$

$$N_\square = \frac{10(10^{-6})(5)(60)}{q\pi\left[(5)(2.54)\right]^2/4} = 1.48 \times 10^{14} \text{ ions/cm}^2$$

For an implantation of 80 keV B$^+$ ions into a silicon target:

$$\bar{R}_p = 2450 \text{ Å}$$

$$\Delta\bar{R}_p = 650\text{Å}$$

$$N_{max} = \frac{0.4N}{\Delta \overline{R}_p} \approx \frac{(0.4)(1.48 \times 10^{14}) \text{ ions/cm}^2}{650 \times 10^{-8}}$$

$$\approx 9.20 \times 10^{18} \text{cm}^{-3}$$

From this information, the theoretical concentration profile can be constructed. It should be noted that this is only a first order approximation to the actual profile. In order to more accurately fit the typical profile achieved in practice which is asymmetrical, higher spatial moments are required. These methods are discussed in much greater detail by Stone and Plunkett (26) and the references therein. The asymmetry achieved in practice has been studied by Schwettman (28) and is shown in Figure 7 from White et al (29).

Masking Techniques

Several techniques can be used for masking the ions so as to define the area to be implanted. Usually for microelectronics a contact mask is required. Typical masks are silicon dioxide (SiO_2), silicon nitride (Si_3N_4), photoresist, or metal films. The basic requirements for a mask include:
 (1) Must be compatible with the photolithographic techniques.
 (2) Pattern definition should be sharp.
 (3) Should have an excellent stopping power.
 (4) Should not contaminate the wafer, and should be easily removable.
    The minimum thickness required for various materials to stop a prescribed percentage of the ions can be estimated by the transmission coefficient, T, given by (26):

$$T = \frac{1}{2} \text{ erfc} \left[\frac{d - \overline{R}_p}{(2)^{1/2} \Delta \overline{R}_p} \right] \tag{7}$$

where d is the mask thickness.
    For large arguments, the complementary error function can be approximated by

$$T \approx \exp(-a^2)/ \ 2(\pi a)^{1/2} \quad \text{where a is given by:} \tag{8}$$

$$a = \frac{d - \overline{R}_p}{(2)^{1/2} \ \Delta \overline{R}_p} \tag{9}$$

To stop 99.99% of the incident ions, $T = 10^{-4}$, yielding

a = 2.8. Hence

$$d_{min} = \overline{R}_p + 3.96 \ \Delta \overline{R}_p \tag{10}$$

The minimum thickness for various masking materials are shown in Figures 8, 9, and 10.

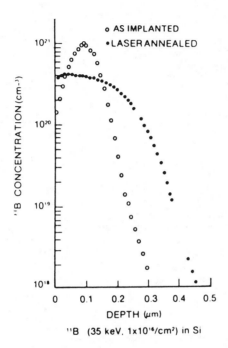

Figure 7. Experimental profile of boron implanted in silicon showing the as-implanted profile and the profile after laser annealing. (Reproduced with permission from Reference 29, copyright 1978, American Institute of Physics).

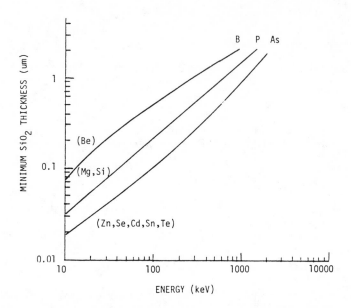

Figure 8. Calculated minimum thickness of SiO_2 for masking B, P, and As at various implantation energies. (Adapted with permission from Reference 27, copyright 1983, John Wiley and Sons).

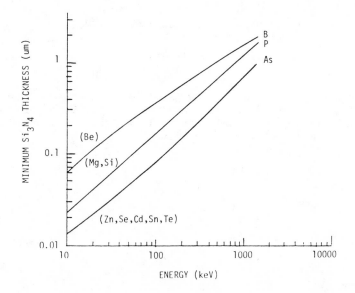

Figure 9. Calculated minimum thickness of Si_3N_4 for masking B, P, and As for various host materials. (Adapted with permission from Reference 27, copyright 1983, John Wiley and Sons).

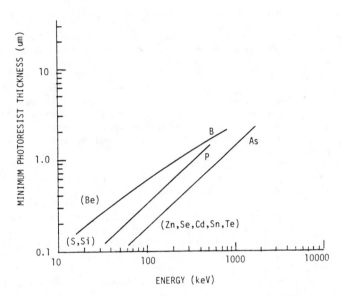

Figure 10.   Calculated minimum thickness of photoresist for masking B, P, and As for various host materials.   (Adapted with permission from Reference 27, copyright 1983, John Wiley and Sons).

Annealing for Electrical Activation

The basic range statistics will provide reasonable predictions of
the unannealed concentration profile. Electrical activation is
necessary to stabilize and ensure that the ions occupy electrically
active sites. There are presently several types of anneal used for
ion-implanted materials, but they can be grouped into two broad
categories: 1) Thermal anneal and 2) laser anneal.

Thermal Anneal. Ion-implanted wafers should be annealed at the lowest temperature possible in order not to diffuse the impurities further into the wafer. At low implanted doses (10^{11} - 10^{12} cm^{-2}) the
electrical activation process can occur at temperatures as low as
260 degrees C in silicon whereas for higher doses (10^{13} - 10^{14} cm^{-2})
570 degrees C may be required. For higher doses (10^{15} cm^{-2})
higher temperatures around 900 degrees C will be required for electrical activation (26). Most anneals are for 30 minutes in a non-
oxidizing ambient. Reference 26 may be consulted for more details
of the thermal anneal.

Laser Anneal. A technique for laser annealing has received considerable attention in recent years. This technique has several
advantages over thermal annealing. Only the surface layer is heated
sufficiently to anneal a shallow layer implant, so the substrate is
unaffected. The heating is also selective, allowing small areas to
be annealed without disturbing other regions. Figures 11, 12, and
13 show typical examples of as-implanted and laser annealed im-
plants (29). The anneals were achieved by using a pulsed ruby laser
with an energy density of 1.65 J/cm^2. Complete electrical activation was achieved.

Annealing in Gallium Arsenide. Gallium arsenide has a greater
variety of defect interactions than silicon. Also, most gallium
arsenide devices are based on majority carrier transport. This
decreases the importance of the minority carrier lifetime. Therefore carrier activation is the primary purpose of the annealing process.

   As-implanted GaAs has no carrier activation, as a rule. Higher
temperatures are required to anneal GaAs material than silicon.
Usually not all carriers are activated even at 900 degrees C. Therefore the annealing of GaAs presents greater problems than silicon.
Moreover, often there is out-diffusion and boundary movement during
anneal, as well as changes in the doping profile.

   Often, GaAs is annealed by using a capping layer during the
anneal. The implant is sometimes made through this cap. The cap
helps to minimize out-diffusion during the anneal. Silicon nitride
and aluminum oxide are typical capping materials. Temperatures as
high as 1100 degrees C have been used for annealing GaAs.

Advances in Device Fabrication by Ion Implantation

The emergence of ion implantation as a primary fabrication process
for a wide variety of devices has occurred at a rapid pace in recent years. The attributes of precise dopant control, low temper-

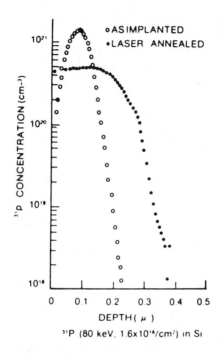

Figure 11. Experimental impurity profile of phosphorus implanted in silicon showing the as-implanted profile and the profile after annealing. (Reproduced with permission from Reference 29, copyright 1978, American Institute of Physics.)

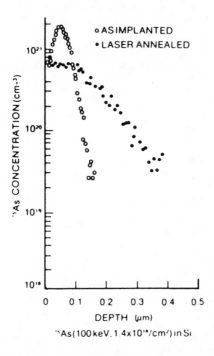

Figure 12. Experimental impurity profile of arsenic implanted
in silicon showing the as-implanted profile and the profile after
laser annealing. (Reproduced with permission from Reference 29,
copyright 1978, American Institute of Physics.)

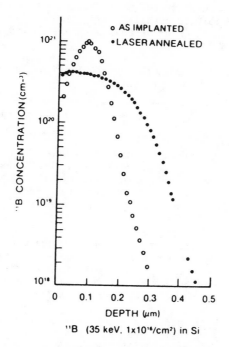

Figure 13. Experimental impurity profile of boron implanted in silicon showing the as-implanted profile and the profile after laser annealing. (Reproduced with permission from Reference 29, copyright 1978, American Institute of Physics.)

ature, and uniformity of impurity concentration has assured its place in the development of new and improved devices. With the advent of laser and thermal pulse annealing, some of the traditional disadvantages of material damage have been minimized.

Several of the present applications of ion implantation to solid state devices are presented in Table I. The tremendous volume of literature prohibits detailed discussion of each application. However, a brief discussion of various applications is presented in the following sections.

Table I.   Summary of Ion Implantation Applications in Device
Fabrication

Silicon Devices	Applications of Ion Implantation
MOS transistor	Ion implantation is used for threshold voltage adjustment. Can also be used for self-alignment with practically no gate overlap capacitance; decouples parasitic capacitances. Can be used to obtain small channel lengths (30).
CMOS transistor	Ion implantation provides threshold adjustment and improves the switching speed of the device. Uniformity and reproducibility of the p-channel is possible; can radiation harden the device (31,32).
Complementary DMOS transistor	Used as a pre-deposition step of the diffusion process. Provides superior delay time (33-35).
ISOMNOSFET	Implantation helps to overcome parasitic capacitance effects by better alignment (36).
Power FET	High energy boron implantation can be used to form highly doped buried layers for vertical channel FETs (37).
Subvolt JFETs	Double implanted subvolt JFETs having supply voltages at 1.5V or lower are possible with ion implantation (38).
Bipolar transistor	Ion implantation allows closely controlled doping profiles and eliminates parasitic devices by better lateral registration. Also can increase cut-off frequency by shallow emitter and base profiles, increasing speed and lowering the noise of the device (39-41).
Integrated injection logic	Shaping of the extrinsic and intrinsic base profiles by ion implantation and by closely

Continued on next page.

Table I.   Continued

Silicon Devices	Applications of Ion Implantation
Integrated injection logic (con'd)	controlled doping density and depth improves the power-delay product and the inverse gain of the vertical npn transistors (42).
P-n diode	Implantation can be used to create very abrupt profiles and to closely control the depth and uniformity of the junction (26).
PIN diode	Ion implantation is used to fabricate the shallow high density p^+ region (26).
Avalanche photodiode	Ion implantation improves yield and signal to noise ratio. Also provides larger minority carrier lifetimes (26).
IMPATT diode	Implantation produces reproducible and narrow base regions for high frequency operation (26).
Solar cell	By the implantation process, solar cells can be made in high resistivity p-silicon by implanting phosphorus to produce shallow n^+ or p-layers. More favorable bulk recombination rates of minority carriers in the p-silicon causes higher collection efficiency (43-45).
Varactor diode	By ion implantation, the slope of the capacitance versus applied reverse voltage can be tailored by implanting phosphorus impurity profiles below the Schottky-barrier (46).
Schottky-barrier diode	Ion implantation is used to control the barrier height. For the PtSi Schottky-barrier diode, the aluminum-Si barrier height can be modified, the peak of which is located in the immediate vicinity of the metal-Si interface (47).
CCDs	The asymmetry necessary for directionality in the transfer of charge can be achieved by implanting packets of increased doping concentration near one edge of each metalization line, making the implanted region more repulsive to minority carriers at the interface than the unimplanted region. Creates a high-low junction effect (26).

Continued on next page.

Table I. Continued

Silicon Devices	Applications of Ion Implantation
Resistors	Ion implantation produces accurate high ohmic resistors on the order of a megohm by controlling the implant fluence and the anneal temperature (26).
HVSOS/MOS	Implantation provides compatibility with CMOS processing and provides low leakage, good gain, low threshold, and high break-down (26).

Compound Semicon-ductor Devices	Applications of Ion Implantation
GaAs MESFETs	Ion implantation process creates low noise, fast devices suitable for microwave or high speed logic devices, with high gain (48-51).
InP MESFETs	These devices can be fabricated with even lower noise and higher gain than in GaAs by ion implantation (52).
GaAs MMICs	Monolithic microwave integrated circuits have been fabricated using ion implantation. This renders feasible the fabrication of monolithic phased array radars. These circuits incorporate active devices, RF circuitry, and bypass capacitors (53-54).
GaAs Complimentary JFETs	By ion implantation, a GaAs enhancement mode JFET has been developed in parallel with the GaAs Schottky-barrier FET or MESFET. It is useful in FET logic DCFL). Creates an ultra low power device with applications to the static RAM (55).
InP FETs	Ion implantation has been successfully applied to fabricate FETs on semi-insulating InP substrate. In FETs with a noise figure as low as 3.5 dB at 12 GHz have been fabricated (52).
GaAs Hall Devices	Implantation has been used to fabricate highly linear GaAs Hall devices. Linearity error is better than $\pm$ 0.03% (56).

The extent of the applications of ion implantation to the fabrication of silicon devices prohibits an in-depth treatment of each application. A more complete discussion of these applications can be found in review form in Reference 26 and the references therein. This presentation therefore deals primarily with the fundamentals of ion implantation and the methodology of applying some of these fundamentals to advanced device technology.

Field Effect Transistor Applications. Among the advantages of ion implantation in field effect device fabrication are:
    (1) Threshold adjustment for compatibility with TTL logic.
    (2) Depletion mode/enhancement mode fabrication.
    (3) Improved frequency response.
    (4) Low temperature and simple process.
    (5) Self-alignment of MOS divices.
    (6) Channel stop fabrication.
    Ion implantation can be used for gate self-alignment in MOS devices. Using boron implantation at about 80 keV, the aluminum gate or thick oxide will mask the ion beam. But when a thin oxide (approximately 1200 Å) is positioned over the remaining part of the silicon surface, the beam will penetrate the thin oxide, and be deposited in the silicon surface at the $Si-SiO_2$ interface, extending the source and drain to the gate boundary. The gate field plate has masked the ions from the gate region. After implantation of about 10^{14} ions/cm^2, the devices are annealed at about 400 to 500 degrees C for about 30 minutes. Figure 14 shows a MOSFET device requiring two implantations, the n-implantation forms a self-aligned source and drain region. The p-implantation permits a short channel length with the high resistivity substrate material required to reduce capacitances.
    Use of this self-alignment process provides the advantages of essentially no overlap capacitance, and improves the speed by a factor of 30 to 40%.
    Ion implantation through an oxide can also be used to increase the threshold voltage outside the channel region (27). Figure 15 illustrates how a channel stopper of this type can be fabricated by a single unmasked high energy implant. To complete this step, a 200-300 keV boron implant is used if the device is an n-channel MOSFET. Since this implantation penetrates deeply under the thin gate oxide, it does not affect the threshold voltage. However, the regions of thick oxide receive a shallow implant in the p-type field regions. It therefore increases the threshold voltage necessary to create parasitic action.
    Ion implantation when used for channel doping has the following advantages:
    (1) The threshold voltage can be adjusted to accommodate direct interfacing with TTL circuitry.
    (2) A high ratio of field threshold-to-device threshold is obtained.
    (3) Depletion mode load/enhancement mode driver circuitry are achieved on one chip.
    This results in higher circuit density, improvement in the speed/power product, and the device can operate from a single 5 volt supply. In this process, ion implantation is used to dope the MOSFET channel with boron at an energy of about 50 keV. The range of these ions are such that they penetrate the 1200 Å gate oxide

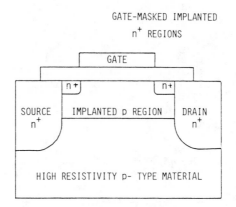

Figure 14.   Typical MOSFET requiring two ion implantations.

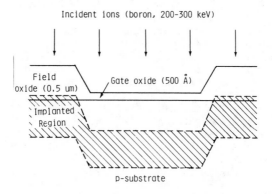

Figure 15.   Ion implantation used for channel stopping.
(Adapted with permission from Reference 27, copyright 1983,
John Wiley and Sons).

and stop in the silicon surface region. Because of ion straggling
in the range, a layer of approximately 0.2 micrometers is obtained
at the silicon surface. The device is then annealed. A PMOS example is presented below:

If 6×10^{16} cm^{-3} boron atoms are implanted in the gate region
of the p-channel MOSFET fabricated on <111> oriented n-type material
with $N_D = 10^{15}$ cm^{-3}, the compensated results yields:

$$P = N_A - N_D = 6 \times 10^{16} - 1 \times 10^{15} = 5.9 \times 10^{16} \text{ cm}^{-3}$$

The channel will be p-type. If the channel is assumed to be uniformly implanted to a depth of 0.2 micrometer, then only 1.18×10^{12}
boron atoms/cm^2 are implanted. For approximately 10^{12}/cm^2 boron
atoms implanted, and if a beam density of 10^{-2} microamperes/cm^2 of
singly charged boron ions is used, then:

$$\text{Implanted atoms/cm}^2 = \frac{\text{(beam current density)}}{q} \times \text{time}$$

$$10^{12}/\text{cm}^2 = \frac{10^{-8} \text{ A/cm}^2}{1.6 \times 10^{-19}\text{C}} \times t$$

Solving for t: t = 22 seconds.

The implantation time required to provide the dose indicated is
then 22 seconds.

In CMOS (complementary metal-oxide-semiconductor) technology,
the p-channel is fabricated by ion implantation. Using this technique, 5% control in the doping is possible and the threshold control is improved.

The ISOMNOSFET has two features which are not in the conventional MNOSFET. It has a stepped oxide in the channel region, and
an ion implanted region adjacent to the channel. By using a
stepped oxide with two oxide thicknesses in the channel, two constant threshold thick-oxide FETs in series with the thin oxide variable threshold device ensures that the composite device maintains
operation in the enhancement mode in both the high and low threshold
states.

Ion implantation is used in the fabrication of low-barrier
PtSi Schottky-barrier diodes. An ion-implanted, shallow n+ layer
has been used by Bindell et. al. (57) to lower the barrier height
of PtSi-n-Si Schottky diodes. Barrier height reductions of up to
200 mV have been achieved. This implant increases the electric
field at the surface, thus lowering the effective barrier height
through an enhanced Schottky lowering effect (57).

Ion implantation has also been used to increase the barrier
height of metal-semiconductor Schottky-barrier diodes (46). This
is accomplished by implanting low energy ions of opposite conductivity type into the semiconductor surface. The implanted ions
change the field and potential in the surface region and reduce
the diode current. Figure 16 shows the variation of current density
versus forward voltage for various values of ion implantation
dose (58).

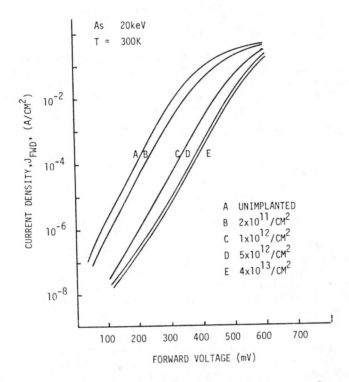

Figure 16.   Calculated diode current density versus forward voltage curves for various values of ion implantation dose. (Adapted from Reference 43.)

In the fabrication of hyperabrupt diodes such as varactors, the flexibility in doping profiles that can be produced by ion implantation allows a wide range of capacitance-voltage characteristics to be designed.

Figure 17 shows the impurity profile of a hyperabrupt Schottky diode. In varactor diodes, ion implantation can be used to tailor the slope of the capacitance versus applied reverse voltage by implanting phosphorus impurity profiles below the Schottky barrier.

Bipolar Transistors and Integrated Injection Logic. The dosage control and profile shaping attributes of ion implantation make it ideal for the fabrication of high frequency, high gain, low noise transistors. Archer (39) in his work separately implanted the inactive and active base regions in 8 GHz transistors. The excellent results were attributable to the close dose control of the active base Gummel number (26). Stone and Plunkett (26) have also found that this technique significantly improves the inverse gain of bipolar transistors when properly carried out. Arsenic emitters are also preferred for low loise and abrupt profiles, thus improving emitter efficiency. Other researchers have used ion implantation in the fabrication of super-gain transistors and high performance transistors with arsenic implanted polysil emitters (40).

Ion implantation is well suited for the design of bipolar structures used in injection logic. Among the attributes of ion implantation for these are 1) ability to fabricate shallow devices thus improving the speed and gain and tailoring the impurity concentration profiles. Separate implantations for the inactive and the active base layers are usually desirable.

It should be remembered that in integrated injection logic, the vertical npn transistor must be operated in the inverse mode. Figure 18 shows the top view of an I^2L unit cell, and Figure 19 shows the cross sectional view. The impurity concentration profile of a typical n^+pnn^+ transistor is shown in Figure 20. Figure 21 shows superimposed on the initial profile, an LEC profile which has been modified to provide an aiding field in the intrinsic base region, plus a low doped region for the inverted collector. This reshaping of the impurity profile has been found to improve the inverse gain and switching speed of the integrated injection logic unit cell (41).

For the most part, one may assume that in the inverse mode the opposite to those forward-mode attributes is true. A larger portion of the field is retarding than aiding, and the overall effect is retarding. This increases the base transit time of the inverse transistor, thus lowering the f_T and switching speed.

It is obvious therefore that if the base profile could be reshaped or tailored to some optimum shape, the switching speed of the transistor would be improved. Several researchers have studied improved shapes of the base profile in search of an optimum. It turns out that optimum shapes for some attributes of the transistor do not optimize others. Among the shapes that have been studied are 1) Gaussian, 2) Complementary error function, 3) Exponential, 4) Parabolic, and 5) Segmented.

The exponential distribution has been shown to be the best of the group for minimum transit time (41). The segmented profile,

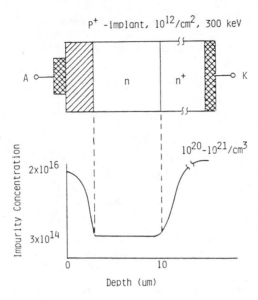

Figure 17.   Hyperabrupt Schottky diode.

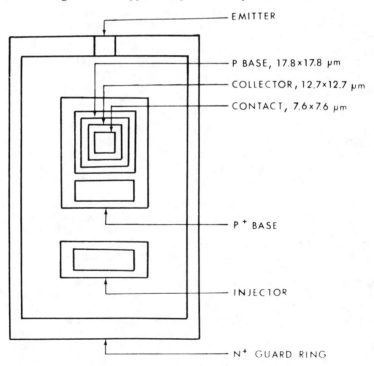

Figure 18.   Top view of integrated injection logic unit cell.
(Reproduced from Reference 41).

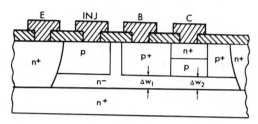

Figure 19.   Cross section view of integrated injection logic chip.   (Reproduced from Reference 41).

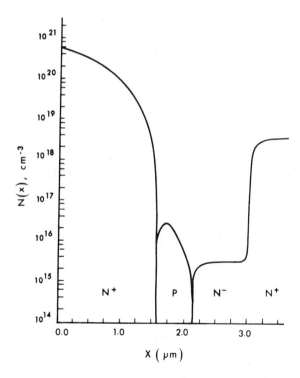

Figure 20.   Impurity concentration profile of a bipolar transistor fabricated by diffusion.   (Reproduced from Reference 41).

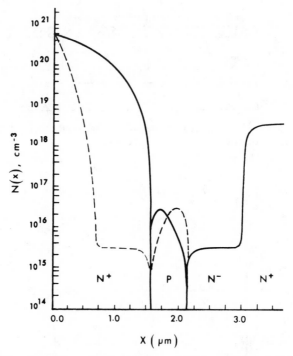

Figure 21.   Impurity concentration profile showing possible
profile shaping by ion implantation for improved operation
in the inverse mode.   (Reproduced from Reference 41).

having small widths of retarding field followed by a large region of
aiding field yields a larger figure of merit for switching transistors (41). The figure of merit used by Maheshwari and Ramanan
is (59):

$$F = \frac{1}{t_b r_b C_c} \qquad (11)$$

where t_b is the base transit time, r_b is the base spreading
resistance, and C_c is the base-to-collector capacitance.
They found that a profile which has a retarding field over a portion of the base and an aiding field over the other improves the
figure of merit compared with the exponential distribution. Ion
implantation could be used to shape this profile. By using ion
implantation, the base Gummel number can be precisely controlled
since the use of current integration for the dose measurement
allows the base impurities to literally be counted. Further discussion of the device parameter optimization by ion implantation
is treated by Stone and Plunkett in Reference 26.
    Experimental measurement of the doping profiles for ion im-
plantation can be performed by anodization and stripping for high
accuracy as given in Reference 41. Figure 22 shows the set-up.
One excellent chemical for the anodization solution is a mixture
of tetrahydrofurfuryl alcohol (THF) and potassium nitrite (KNO_2).
The proper mixture is 2.8 g of KNO_2 per 100 ml of THF. The solution should be irradiated during the process with a tungsten-hal-
ogen lamp to accelerate the anodization process. A typical constant
current used is about 10-15 mA/cm^2. The oxide thickness is proportional to the final forming voltage, and for the conditions described is about 4 Å/volt (41). Figure 23 shows a typical measured
and plotted profile. An algorithm for processing the data is given
in Reference 41.

Ion Implanted Resistors. If a substrate is doped with a layer of
impurities opposite to that of the background doping, the resistivity
is given by the total number, N_s, of the mobile carriers/cm^2 and the
mobility:

$$R_s = \frac{1}{q\mu N_s} \qquad (12)$$

Ion implantation can provide accurate control over N_s and reduce
the mobility (26).

Advances in Compound Semiconductor Ion-implanted Devices

Many of the device technologies which are common in silicon,
such as MOS technology, have yet to be developed in the compound
semiconductors. Researchers are presently pursuing these areas.
A variety of problems are presented in the development of some of
these technologies. Heavy doping necessary for certain areas of the
devices is difficult to achieve by simple processes in gallium arsenide. In spite of these shortcomings, there are several areas in
which ion implantation has already been successfully used. With

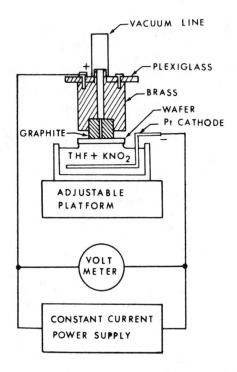

Figure 22. Experimental laboratory apparatus for the anodization and stripping of silicon. (Reproduced from Reference 41).

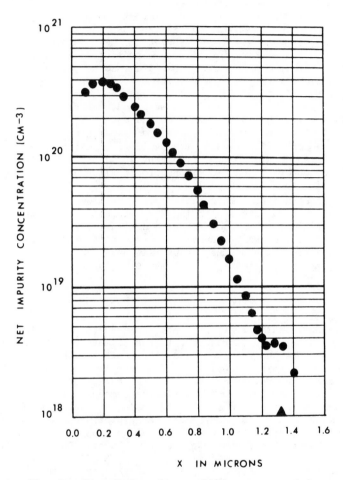

Figure 23. Results of impurity profile measurement by anodization and stripping technique. (Reproduced from Reference 41).

increasing needs for development in these areas of Gigabit logic
and microwave power devices, compound semiconductors such as gallium arsenide are excellent candidates. Gallium arsenide shows
significant promise because of its higher mobility and other attributes. A brief discussion of several applications of ion implantation in gallium arsenide devices are outlined here. Golio
(60) et. al. have studied the potential of a number of compound
semiconductors for low noise microwave MESFET applications. The
industry is making continuous advances in the development of GaAs
field-effect transistors for microwave applications. These devices have been shown to exhibit excellent noise performance through
the K-band. Moreover, as the requirements for VLSI memory and logic
become more stringent, and greater demands are placed upon the attributes of speed and low power consumption, MESFETs look more and
more attractive. Golio's study included models of GaAs, InP, GaInAs,
InPAs, and other devices. Comparison of a $Ga_{0.5}In_{0.5}As_{0.96}Sb_{0.04}$
device to a conventional gallium arsenide device indicated that a
decrease in the minimum noise figure by a factor of two is possible.
Some of the other compound semiconductor devices also showed improvement in the noise figure.

Sleger et al. (52) have reported the development of ion-implanted one micron gate length InPFETs with noise figures as low as
3.5 dB at 12 GHz. When compared with similar GaAs devices, superior gains at microwave frequencies over that of GaAs was demonstrated. Further improvement in profile optimization of the implanted area appears feasible as a means of further reduction in
the noise figure.

The development of monolithic microwave amplifiers formed by
ion implantation into LEC gallium substrates have been reported by
Driver et al. (54). These monolithic power integrated circuits on
semi-insulating gallium arsenide are expected to provide improved
bandwidth over hybrid microwave circuits because of improved matching environments and the reduction of parasitic inductances due to
wire bonds. As this technology is evolved, it is entirely feasible
that phased array radars could be fabricated with these monolithic
chips. Driver has built one and two stage monolithic GaAs power
amplifiers operating from 5 to 10 GHz using standard photolithographic masking with the addition of a chlorobenzene process to
help the metal liftoff. Powers of 28 dBm across a bandwidth of
5.7 to 11 GHz with 6 dB gain have been achieved. Gupta et al. (51)
have also reported process improvements in MESFET GaAs monolithic
microwave circuits. Gupta (52) in a later paper has detailed an
ion-implantation fabrication process for fabricating GaAs monolithic microwave integtated circuits (MMICs), incorporating active
devices, RF circuitry, and bypass capacitors.

Feng et al. (49) have reported the fabrication of GaAs MESFETs
by ion implantation into MOCVD (metal organic chemical vapor deposition) buffer layers. Recently, metal organic chemical vapor
deposition has been used for the fabrication of the channel layers
in GaAs FETs (49). Feng reports a reporducible process using ion-
implanted MOCVD buffer layers for low noise MESFETs. The buffer
layers were grown by the MOCVD technique on <100> Cr-doped semi-
insulating GaAs substrates. After degreasing, the substrates were
etched in HCL for one minute and in a solution of 5:1:1 H_2SO_4:

H_2O_2: H_2O at 40 degrees C for two minutes; then they were rinsed in deionized water for 20 minutes. After blowing dry in dry N_2 and loaded into the MOCVD chamber, the growth was done at a temperature of 650 degrees C at atmospheric pressure. The buffer layer is approximately 2 microns thick with net carrier concentration of $3 \times 10^{15} cm^{-3}$ and a mobility of 6500 cm^2/V-sec.

The channel layer was formed by direct ion implantation with a Si ion dose of $6.5 \times 10^{12} cm^{-2}$ at 100 keV. Annealing at 850 degrees C in H_2-As_4 was performed. Results of the best device showed a noise figure of 1.46 dB with a gain of 10.2 dB at 12 GHz. The results indicate that a high degree of microwave uniformity can be achieved by ion implantation into MOCVD buffer layers.

Zuleeg et al. (55) have reported the fabrication of a double-implanted GaAs complementary JFET. The GaAs enhancement mode JFET was developed in parallel with the GaAs Schottky-barrier FET or MESFET. It is useful for direct coupled FET logic (DCFL). Reduction in the required power level by an order of magnitude (from about 100 microwatt/gate) is possible by using complementary n-channel and p-channel enhancement mode JFETs. Zuleeg has reported a double-implantation of n-channel and p-channel enhancement GaAs FET and its application to an ultra-low power static RAM. The reader is referred to Zuleeg's paper for the process steps.

Heterojunction bipolar transistors (HBTs) are currently receiving increased attention by researchers for high speed applications. To meet these requirements, transistors must be fabricated with smaller emitter widths and better contacts to the p-type base region. Ion implantation is presently being used to accomplish these requirements. The high temperature annealing required presents a problem, however. Asbeck et al. (61) have reported a thermal annealing (pulsed) technique for annealing the ion implanted devices. Application of the technique was made to the fabrication of Be-implanted MBE (molecular beam epitaxy) grown GaAlAs/GaAs heterojunction bipolar transistors. The thermal annealing was shown to compare well with the furnace annealing (non-pulsed) without the occurrence of impurity diffusion.

Systems for Ion Implantation

Most of the ion implantation systems have the same basic elements, only differing in the details and perhaps level of automation. A schematic diagram of a typical ion implantation system is shown in Figure 24.

For the purpose of this discussion it will be assumed that the Nielson source is used. It consists of a cylindrical arrangement of a tungsten helical cathode , a cylindrical anode of graphite and a magnet coil. Solid materials can be vaporized in an oven with two types of crucibles where temperatures ranging from 170 to 900 degrees C can be covered. The lifetime of a source is about 30 hours. Typical source materials are BF_3, phosphorus (PF_3), and AsF_3.

Typical ion sources work on the basic principle that a confined electric discharge or arc is partially or completely sustained by the gas or vapor of the material that is being ionized. The hot cathode source consists of a hot emitting electrons (usually

a filament) and an anode. The presence of a small amount of gas
whose ionization potential is less than the potential between the
electrodes will cause the primary electrons to produce positive ions.
A diagram of a Nielson source is shown in Figure 25. For further
discussion of ion sources the reader is directed to Reference 62.
    The beam is then focused, pre-accelerated, and passed through
a mass analysis stage. After fucusing, the beam is accelerated to
its final energy by a linear accelerator. The beam is again focused by a doublet quadrupole before entering the x-y scanning system. Electrostatic scanning in the x and y directions is provided.
A trap is provided for uncharged, neutral species while the selected ions are deflected to the target. A vacuum of below 10^{-6}
Torr is necessary for the system.
    A typical target chamber is shown in Figure 26. The chamber
can be a carousel type which rotates to a series of wafers for
implantation.

Appendix

Chemical Processing for Ion-implanted Integrated Circuits

Detailed cleaning and processing steps are often not published
by the industry. For university researchers and students, however,
this often becomes a problem. Often they are forced to consume an
inordinate amount of time developing a routine process for cleaning
and basic processing. The author faced such a delimma a few years
ago while a doctoral student, and in order to ease the burden of
some poor graduate student, the following procedure has been found
acceptable by the author for certain photolithographic operations.
It is hoped that this may save some time for the student if he
does not have a workable procedure.

Initial Wafer Cleanup. The following procedure should be done in
a class 100 clean room:
1.   10% HF in DI water solution at 25 deg. C for 6 min.
2.   Cascade rinse in DI water for 6 min.
3.   Tetrachlorethylene at 80 deg. C for 6 min.
4.   Cascade rinse in DI water for 6 min.
5.   Ammonium Hydroxide mixture at 80 deg. C for 6 min.
6.   Cascade rinse in DI water at 25 deg. C for 15 min.
7.   HCL mixture at 25 deg. C for 6 min.
8.   Cascade rinse in DI water at 25 deg. C for 15 min.
9.   Blow dry with dry nitrogen.
10. Bake in process clean oven at 135 deg. C for 10 min.
11. The wafer should be used within 20 minutes after the cleanup.

Predeposition Cleanup. The following procedure should be done in
a class 100 clean room:
1.   Agitate in Nitric acid at 90 deg. C for 6 min.
2.   Cascade rinse in DI water for 10 min.
3.   Slowly agitate in 10% HF for about 5 sec. depending on doping
     level of prior steps.
4.   Rinse immediately in DI water for 15 min.
5.   Cascade rinse in DI water for 15 min.
6.   Blow dry with dry nitrogen.

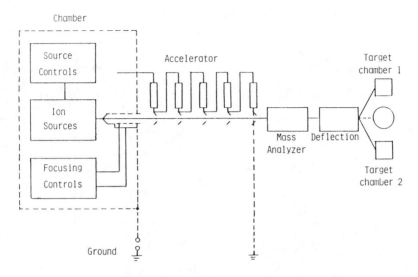

Figure 24. Block diagram of an ion implantation system.

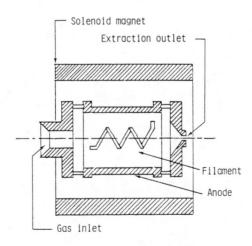

Figure 25. Nielson-type source for ion implantation.

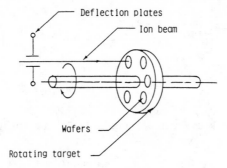

Figure 26.   Target of an ion-implantation system.

7. Bake at 135 deg. C for 10 min.

Oxide Removal. The following procedure should be done in a class
100 clean room:
1. Use Bell 2 etchant (there is less undercutting than with HF).
2. The etch rate is approximately 750 Angstroms/min. in low
   doped oxides.
3. Solution preparation:
   a. 200 ml DI water
   b. 200 g. NH_4F (filtered Ammonium Flouride)
   c. 45 ml HF (49%)

Aluminum Etching Procedure. The following procedure should be
done in a class 100 clean room:
1. Etch at room temperature. The following mixture is satisfactory:
   a. 75% H_3PO_4
   b. 22% Acetic acid
   c. 3% Nitric acid
2. Agitate constantly and observe wafer to see when the slice
   is clear. Etch rate is approximately 1 micron per 30 minutes.

Formulas for Mixtures Used in Initial Wafer Cleanup.
1. Ammonium Hydroxide mixture: Mix 5 parts DI water, one part
   NH_4OH; heat to 80 deg. C; then add 1 part H_2O_2 (30% unstabilized)
   This is added just before using and heated back to 80 deg. C.
2. HCL Mixture: Mix 6 parts DI water, 1 part HCL; heat to 80
   deg. C; add 1 part H_2O_2.

Literature Cited

1. Agajanian, A. H. Rad. Eff. 1974, 23, 73.
2. Brewer, G. R. IEEE Spect. 1971, 8, 23.
3. Dearnaley, G. "Ion Implantation, Annual Review of Materials
   Science." 1974, 4, 93.
4. Dearnaley, G.; Freeman, J. H.; Nelson, R. S.; Stephen, J.
   "Ion Implantation Defects in Crystalline Solids"; North
   Holland, Amsterdam, 1973, 8.
5. Dill, H. G.; Finnila, R. M.; Luepp, A. M.; Toombs, T. N. Solid
   St. Tech. 1972, 15, 27.
6. Gibbons, J. F. Proc. IEEE, 1968, 56, 295.
7. Gibbons, J. F. Proc. IEEE, 1972, 60, 1062.
8. Lee, D. H.; Mayer, J. W. Proc. IEEE, 1974, 62, 1241.
9. Mayer, J. W.; Marsh, O. J. "Ion Implantation in Semiconductors",
   Academic Press, New York, 1971, 1.
10. Mayer, J. W., Davies, J. A.; Eriksson, L. "Ion Implantation
    in Semiconductors", Academic Press, 1970.
11. Mazzio, J., "NTIS Report No. SC-B-710148", Sandia Laboratories,
    1971.
12. Morgan, R.; Greenhalgh, K. R., "AERE-Bib-176", Atomic Research
    Establishment, Harwell, England, 1972.
13. Pickar, K. A., "Ion Implantation in Silicon--Physics, Pro-
    cessing, and Microelectronic Devices", Academic Press, New
    York, 1975, 5.

14. Plunkett, J. C.; Stone, J. L. Solid St. Tech. 1975, 18, 49.
15. Reddi, V. G. K.; Yu, A. Y. C. Solid St. Tech. 1972, 15, 35.
16. Seager, D. K.,"NTIS Report No. SC-B-71048, Supplement I",
    Sandia Laboratories, 1973.
17. Stephens, J., The Radio and Elect. Engr. 1972, 42, 265.
18. Stone, J. L.; Plunkett, J. C. Solid St. Tech. 1976, 19, 35.
19. Stone, J. L.; Plunkett, J. C. "Ion Implantation Processes
    in Silicon, Ch 2, Impurity Processes in Silicon", North
    Holland, Amsterdam, 1981.
20. Titov, V. V. Phys. Stat. Sol. 1978, 48, 13.
21. Lindhard, J.; Scharff, M.; Schiott, H. E. Mat. Fys. Medd.
    Dan. Vid. Selsk 1963, 33, 14.
22. Gibbons, J. F. Proc. IEEE. 1968, 56, 295.
23. Johnson, W. S.; Gibbons, J. F. "Projected Range Statistics
    in Semiconductors", Standord University, 1969.
24. Lindhard, J.; Scharff, M. Phys. Rev. 1961, 124, 128.
25. Gibbons, J. F.; Johnson, W. S.; Mylroie, S. W. "Projected
    Range Statistics," 2nd Ed., Dowden, Hutchinson, and Ross,
    1975.
26. Stone, J. L.; Plunkett, J. C. "Ion Implantation Processes
    in Silicon, Ch 2 Impurity Doping Processes in Silicon",
    Edited by F. Y. Y. Wang, North Holland, Amsterdam, 1981.
27. Ghandhi, S. K. "VLSI Fabrication Principles", Wiley & Sons,
    1983.
28. Schwettman, F. N. J. Appl. Phys. 1974, 45, 1918.
29. White, C. W.; Christie, W. H.; Appleton, B. R.; Wilson, S. R.;
    Pronko, P. D.; Magee, C. W. Appl.Phis Lett. 1978, 33, 662.
30. MacDougall, J. K. Electronics. 1970, 43, 86.
31. Dingwall, A. G. F.; Stricker, R. E. IEEE J. Solid St. Circ.
    SC-12, 1977, 344.
32. Declercq, M. J.; Laurent, T. IEEE J. Solid St. Circ. 1977,
    SC-11, 264.
33. Lin, H. C.; Halsor, J. L. IEEE J. Solid St. Circ. 1976,
    SC-11, 443.
34. Masuhara, T.; Muller, R. S. IEEE J. Solid St. Circ. 1976,
    SC-11, 453.
35. Troutman, R. R. IEEE Trans. Electron Dev. 1977, ED-24, 182.
36. Krick, P. J. IEEE Trans. Electron Dev. 1975, ED-22.
37. Ozawa, O.; Iwasaki, H. IEEE Trans. Electron Dev. 1978,
    ED-25, 56.
38. Satwinder, et. al., Solid St. Electronics. 1981, 25, 791.
39. Archer, J. A. Solid St. Electronics, 1974, 17, 387.
40. Gregg, W. M.; Saltich, J. S.; Roop, R. M.; George, W. L.
    IEEE J. Sol. St. Circ. 1976, SC-11, 485.
41. Plunkett, J. C. PhD Dissertation, Texas A&M University,
    1978.
42. Young, R. T.; White, C. W.; Narayan, J.; Westbrook, R. D.;
    Wood, R. F.; Cristie, W. H. "13th Annual IEEE Photovoltaic
    Specialist Conf.", Washington, DC, June 5-8, 1978.
43. Kirkpatrick, A. R.; Minnucci, J. A.; Greenwald, A. C.
    "13th Annual IEEE Photovoltaic Specialist Conf." Washington,
    DC (June 5-8, 1978).
44. Pai, Y. P.; Lin, H. G. "13th Annual Photovoltaic Specialist
    Conf." Washington, DC (June 5-8, 1978).

45. Mulder, J. C.; Grob, J. J.; Stuck, R.; Siffert, P. "13th Annual IEEE Photovoltaic Specialist Conf." Washington, DC (June 5-8, 1978).
46. Moline, R. A.; Foxhall, G. IEEE Trans. Electron Dev. 1972, ED-12, 267.
47. Ashok, S. et al. EDL-5. 1984, 2, 48.
48. Masao, Ida, et al. Solid State Electronics. 1981, 24, 1099.
49. Feng, M.; et al. EDL-5. 1984, 1, 18.
50. Taylor, G. et al. IEEE Trans. Electron Dev. 1979, ED-26.
51. Gupta, A. K. et al. IEEE EDL. 1983, 12, 1850.
52. Sleger, K. J., et al. IEEE Trans. Electron Dev. 1981, ED-28, 1031.
53. Gupta, A., et al. IEEE Trans. Electron Dev. 1983, ED-30, 16.
54. Driver, M., et al. IEEE Trans. Electron Dev. 1981, ED-28, 191.
55. Zuleeg, R., et al. IEEE EDL. 1984, EDL-5, 21.
56. Hara, T., et al. IEEE Trans. Electron Dev. 1982, ED-29, 78.
57. Bindall, J. B., et al. IEEE Trans. Electron Dev. 1980, ED-27, 420.
58. Pai, Y. P.; Lin, H. C. Solid State Elect. 1981, 24, 929.
59. Maheshwari, L. K.; Ramann, K. V. Solid State Elect. 1976, 19, 307.
60. Golio, J. M., et al. IEEE Trans. Electron Dev. 1983, ED-30, 1844.
61. Asbeck, P. M., et al. IEEE EDL. 1983, EDL-4, 81.
62. Nielson, K. O. Nucl. Instrum. and Methods. 1957, 1, 289.

RECEIVED August 1, 1985

Plasma-Assisted Processing
The Etching of Polysilicon in a Diatomic Chlorine Discharge

Herbert H. Sawin, Albert D. Richards, and Brian E. Thompson

Department of Chemical Engineering, Massachusetts Institute of Technology, Cambridge, MA 02139

The fundamental kinetics and transport properties of plasma processes are reviewed and applied to polysilicon etching in Cl_2 discharges. The relative neutral flux, ion flux, and ion energy is critical in controlling the directionality of the etching process. The electron density and energy have been estimated from electrical impedance measurements of Cl_2 discharges. With increasing frequency or pressure; the electron density of the Cl_2 discharge increases while the average electron energy decreases. As the electron energy declines, the fraction of energy dissipated in ionizing collision events decreases while the lower energy dissociative processes producing Cl and Cl^- increase. The larger production of Cl, which chemically etches the doped polysilicon, leads to isotropic etching. The ion flux to the sample does not vary directly with the electron density and is believed to be strongly affected by the presence of negative ions.

In this paper the kinetic and transport properties of glow discharges are reviewed and used to describe current work on a particular process, the etching of polysilicon with Cl_2. In plasma processes for the fabrication of microelectronics, radio frequency glow discharges are used to etch, deposit, sputter, or otherwise alter the wafer surfaces. Radio frequency (rf) discharges produce highly reactive neutrals and ions at low tempertures by the introduction of energy into the plasma through its free electrons. A plasma can be defined as a partially ionized gas in which the charged species have a sufficient concentration that they interact significantly through coulombic forces. The charged species within the plasma respond to oppose any applied field and maintain an overall neutrality of the plasma.

A glow discharge is a non-equilibrium plasma in which the electrons have a greater average energy than the ions and neutrals. A plasma is sustained by the introduction of energy from an electric or

0097–6156/85/0290–0164$06.00/0

magnetic field. The electrons are accelerated by the field, thus gaining energy which is passed to the neutrals through collisions. The low rate of energy exchange in elastic collisions between the much heavier neutrals and the electrons results in the significantly higher average energy of the electrons. The electrons, therefore, can have sufficient energy to produce (through inelastic collisions) significant amounts of ions, free radicals, and other excited species without appreciably heating the gas. In this manner, concentrations of free radicals and ions which would normally be created at only at flame temperatures, can be maintained at room temperature. Since the surface and gas-phase reactions of free radicals and excited species have lower activation energies than similar thermally induced reactions, the kinetics can be greatly enhanced.

Plasmas are used in three major microelectronics processes; sputtering, plasma enhanced chemical vapor deposition (PECVD), and plasma etching. In each, the plasma is used as a source of ions and/or reactive neutrals and is sustained in a reactor so as to control the flux of neutrals and ions to a surface. The typical ranges of properties for a glow discharge used in microelectronic fabrication are as shown in Table I.

In sputtering, ions are extracted from a plasma, accelerated by an electric field, and bombarded upon a target electrode composed of the material to be deposited(1). The bombarding ions dissipate their energy by sputtering processes in which the surface atoms are ejected primarily by momentum transfer in collision cascades. The ejected atoms are deposited upon wafers which are placed within line-of-sight of the target electrode, thus inducing the vapor transport of material without appreciably heating either the target electrode or the wafers on which the film is deposited.

Plasma enhanced chemical vapor deposition uses a discharge to reduce the temperature at which films can be deposited from gaseous reactants by creating free radicals and other excited species which react within the gas-phase and on the surface(1). During processing, the surface is typically heated to acquire a better quality of deposition film. Also, the ion flux from the plasma can be used to clean the surface before the deposition begins to improve the film. The ion flux is also thought to alter the film during deposition.

In plasma etching, the plasma produces both highly reactive neutrals (e.g. atomic fluorine) and ions which bombard the surface being etched(2). The neutrals react with the surface to produce volatile species which desorb and are pumped away. Ion bombardment often increases the etching rate by removing surface contaminants which block the etching or by directly enhancing the kinetics of the etching. Very large scale integration (VLSI) requires the etching of films which have thicknesses that are comparable to the lateral feature dimensions. Plasma etching processes are required to pattern such features since they are capable of necessary anisotropy while wet etching processes (which use aqueous acids or bases) are typically isotropic, producing undercutting of the pattern at least equal to the film thickness. Figure 1 shows the etching profiles produced by isotropic and anisotropic etching processes. An electric field is formed by the contact of a plasma with a surface creating a plasma sheath in which the ions are accelerated along the macroscopic surface normal. Due to this directionality, a larger flux of ions

Table I.   Typical Property Ranges of Glow Discharges Used in
           Microelectronic Fabrication

Property	Range
Pressure	0.01 to 1 torr
Electron density	10^8 to 10^{11} cm^{-3}
Average electron energy	1 to 10 eV
Average neutral and ion energy	0.025 to 0.035 eV
Ionized fraction of gas	10^{-5} to 10^{-7}
Neutral diffusivity	100 to 10,000 cm^2/s
Free radical density	less than 30%
Power dissipation	0.1 to 1 W/cm^2

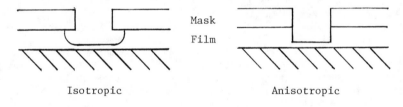

Mask

Film

Isotropic   Anisotropic

Figure 1.   Profiles produced by isotropic and anisotropic etching.

impinges on those surface features which are parallel with the macroscopic surface.

Plasma etching processes are typically broken into two categories, "plasma etching" and "reactive ion etching" (RIE). "Plasma etching" processes are typically distinguished as having higher pressures (0.1 to 1 torr) and either having symmetric (equal area) electrodes or the plasma isolated from the wafers. RIE processing is performed at lower pressures (0.01 to 0.1 torr) and in an asymmetric reactor. In RIE, the wafers are placed on the smaller electrode which increases the energy of the ions striking the surface. Since the power can be increased to produce anisotropic etching in a symmetric reactor, a better distinction would be according to the dominant etching processes; chemical etching by neutrals which etch isotropically ("plasma etching") or ion-induced etching which yields anisotropic profiles (RIE).

Plasma Kinetics

The kinetics which are characteristic of glow discharges are a result of collisions between energetic electrons and neutral gases. To calculate the rate of production by electron collisions, the electron energy distribution $f(\varepsilon)$ must be known. It can be calculated using the Boltzmann transport equation, a differential concentration balance for electrons which considers the movement of the electrons and the energy distribution caused by their coupling with fields and their collisions with other particles($\underline{3}$). The energy distribution is a function of both the applied field and the composition of the gas with which the electrons collide. Assuming that 1) the cross-section for elastic collisions is inversely proportional to the electron velocity, 2) elastic collisions are the dominant mode of energy loss, 3) the electron energy relaxation time is much greater than the field frequency, and 4) the collision frequency is much greater than the field frequency; the time-averaged portion of the distribution function is Maxwellian, where the electron temperature T_e is as below,

$$f(\varepsilon) \;\alpha\; \sqrt{\frac{\varepsilon}{(k_B T_e)^3}} \;\; \exp\left[-\frac{\varepsilon}{k_B T_e}\right]$$

where

$$k_B T_e = \frac{e^2 E_0^2 M}{6 m_e^2 \nu^2} \;\;\alpha\;\; \left[\frac{E_0}{p}\right]^2$$

Here, ε is the electron energy, k_B is the Boltzmann constant, E_0 is the effective electric field, ν is the elastic collision frequency, M is the mass of the gas particle, m_e is the electron mass, and p is the pressure. Although these assumptions are not entirely correct, a Maxwellian distribution function is often used for the calculation of plasma kinetics due to its simplicity. The electron temperature is proportional to the square of E_0/p, which can be physically interpreted as being proportional to the average amount of energy a electron acquires between collisions. In typical processes, however, inelastic collisions are the dominant mode of energy loss which results in experimentally observed average electron energies which scale roughly as the square root of E_0/p (3). In all cases the average electron energy is monotonically related to E_0/p.

The rate of production of the ith species within the plasma by electron impact with its percursor n_i can be calculated using the rate coefficient k_i;

$$r_i = k_i n_e n_i$$

where k_i is calculated by the integration over all electron energies of the reaction probability for collisions between electrons and n_i;

$$k_i = \int_0^\infty \sqrt{\frac{2\varepsilon}{m_e}} \; \sigma_i(\varepsilon) f(\varepsilon) d\varepsilon$$

where the first term in the integral is the electron velocity and $\sigma_i(\varepsilon)$ is the collision cross-section as a function of electron energy. Therefore, if the electron energy distribution function and the collision cross-section are known, the rate for the generation of species within a plasma can be calculated as second-order gas-phase reactions. However, the cross-sections are typically not available and the electron distribution function is not known, making the above analysis useful in understanding the plasma processes, but not yet used in the design of plasma reactors.

Transport of Charged Species

Neutral species within a plasma undergo diffusive and convective transport identical to that in a gas, however, charged species have an additional restriction as a result of the conservation of charge neutrality over distances which are greater than the Debye length of the plasma (of the order of 100 µm). Therefore, in a plasma which contains an equal number of positive ions and electrons, their transport is coupled through the coulombic interactions forcing them to have an equal diffusivity called the ambipolar diffusivity, D_a (3). The fluxes Γ_i's within a plasma of charged species in terms of the densities n_i's, diffusitivies D_i's, and mobilities μ_i's are

$$\Gamma_e = -D_e \nabla n_e - n_e \mu_e E_{sc}$$

$$\Gamma_+ = -D_+ \nabla n_+ + n_+ \mu_+ E_{sc}$$

where E_{sc} is the space charge field induced by the charge separation. Assuming $n_+ = n_e$, the above expressions can be solved yielding

$$\Gamma = - \frac{D_+ \mu_e + D_e \mu_+}{\mu_e + \mu_+} \nabla n = - D_a \nabla n$$

Using the fact that $\mu_e \gg \mu_+$ and the Einstein relation between diffusivity and mobility ($D/\mu = k_B T/e$), the ambipolar diffusivity can be simplified to

$$D_a \approx D_+ \left[1 + \frac{T_e}{T_+} \right] \approx D_+ \frac{T_e}{T_+}$$

Therefore, the diffusivity of ions is enhanced by the factor T_e/T_+
(≈100) while that of electrons is reduced.

A surface which is in contact with a plasma, but which is electrically isolated, must reach a steady-state potential with time such
that the fluxes of positive and negative charge to the surface are
equal. Under such conditions, the surface potential is said to be at
the floating potential and is negative with respect to the plasma
which reduces the flux of electrons but enhances the flux of positive
ions providing a net neutral flux. This difference is manifested in
an electric field which is analogous to the space charge field within
the bulk of the plasma which enforces the equality of diffusive
fluxes. Since the electric field points into the surface, a sheath
which is depleted in electrons is formed at the surface. Simplisti-
cally, the ions diffuse into the field formed at the surface and are
captured. Therefore, the flux of positive ions is equal to that of
the electrons to a floating surface and is determined by the rate of
ion diffusion to the sheath(4). Due to the coupling of the positive
ions with the electrons, the rate of diffusion is enhanced by the
higher energy of the electrons and the flux to the sheath is given by

$$\Gamma_+ = 0.6n_+\sqrt{\frac{k_B T_e}{m_+}}$$

If the potential of a surface which is in contact with a plasma is
positive with respect to the floating potential, a large increase in
the electron flux occurs which opposes the change in potential.
Similarly, a surface with a potential negative with respect to the
floating potential receives a greatly reduced flux of electrons. The
flux of ions is limited by the diffusion of ions to the sheath
region, but is not influenced greatly by variations of electrode
potential from that of the floating potential. The primary response
is to enhance or reduce the electron flux to the surface due to the
higher mobility of the electrons. Because of the large electron flux
induced by positive biasing of the surface but relatively small
variations in ion fluxes with negative biases, the plasma sheath acts
electrically as a leaky diode.

In rf discharges, the diode nature of the sheath can induce
rectification of the rf signal causing the average sheath potential
to approach one quarter of the peak-to-peak voltage V_{pp} in a symmetric electrode reactor, while voltages in an assymetric reactor can
approach $V_{pp}/2$ at the smaller electrode(4). Ion bombardment energies
can, therefore, vary from a few eV to over 1000 eV. However, ion-
neutral collisions within the sheath can cause significant reductions
in the energy of most ions striking the surface from energies calculated using the sheath potential. Also, the higher sheath potentials
induce a greater charge separation and sheath thickness.

Surface Kinetics

The production of free radicals and other excited neutral species
within the plasma can enhance the rate of reaction in the gas and
with the surface by reducing the activation energy for reaction
and/or dissociative adsorption on the surface. For example, CF_4 does
not adsorb on silicon, however, both of the major plasma products

(CF_3 and F) have high sticking probabilities and chemisorb with low activation energies(5). In addition, when the surface reaction is rate limiting, a sensitivity to wafer temperature is often observed. The chemical nature of free radical reactions permits the choice of gases which selectively etch the desired films, but not other films.

Ion bombardment can both induce sputtering and enhance chemical processes. Ions bombarding the surface with an energies greater than approximately 20 eV can sputter material from the surface as described previously for the sputtering process. This etching process is physical in nature and is not very selective to the composition of the surface. In plasma etching processes, the simultaneous flux of neutrals and ions can have a strong synergistic effect in which the rate of etching by the combination is significantly higher than the sum of individual processes. Enhancements of greater than 25 have been found in the fluorine-silicon system(6). The ion-enhancement mechanism is not yet clearly understood, but it is believed to be a result of increased sputtering yields with the weakening of surface bonds caused by the halogenation of the surface. There is evidence that halogenation of the surface extends a number of atomic layers into the surface(7). Ion bombardment can also induce directionality by removing species from the surface which either block the neutral species from reacting with the surface or preferentially react with the neutral etching species reducing the neutral etching. If the ion-enhanced etching processes are rapid with respect to the neutral etching processes, the etching process is anisotropic.

Reactor Design

Plasma reactors are designed to control both neutral and ion flux to the wafers being processed, and thus, the process kinetics. Many different designs have been developed, but they can be categorized into two major types: volume loading and surface loading(7). In a volume loading reactor, the wafers are stacked with a gas space between wafers. The term volume loading refers to the number of wafers being processed as proportional to the volume of the reactor. The plasma is created near the wafers and the neutral reactants diffuse from the discharge, between the wafers, and to the wafer surfaces. The flux of ions to the wafer surfaces is minimized by this design, thus, the process is chemical and isotropic.

In the surface loading reactors, the wafers are placed on electrode surfaces, therefore, the number of wafers is limited by the available electrode area. The discharge is created between the electrodes and contacts the wafers. The wafers receive both appreciable ion and neutral fluxes from the plasma. If the area of one electrode is smaller than that of the opposing electrode, the rectifying behavior of the sheath formed at the surface causes the smaller electrode to be bombarded by higher energy ions. By inserting additional electrodes and/or altering the geometry, the ion flux to the wafer can be controlled somewhat independently of the neutral flux.

Plasma Impedance Analysis

To analyze the processes within a discharge, knowledge of the electron density and energy distribution is needed. Langmuir probes have been successfully used in DC discharges but are not easily applied to the rf discharges because of the rapid potential oscillations(9). An estimate of the electron density and the average electric field within a discharge can made from the measurement of the electrical impedance of the plasma. Researchers have measured the impedance by a number of techniques, including the analysis of the impedance matching network used in coupling the power to the discharge(10).

In this study, the impedance of the discharge was measured by acquiring the voltage and current waveforms to the powered electrode. Consider a parallel plate reactor in which the plasma is confined between the electrodes as shown in Figure 2. At frequencies greater than 5 MHz, the electrical response is primarily determined by the electrons since the larger mass of the ions limit their response. The plasma sheaths are modeled as capacitors since they have a low con-centration of electrons due to the formation of the sheath electric field which repels the electrons. The bulk of the plasma acts as a resistor where the resistance is determined by the concentration and mobility of the electrons. Using the approximation that the electron density is constant throughout, the plasma impedance is modeled as a resistance R and the plasma sheath as a capacitance C_s. Therefore,

$$R = \frac{|V|}{|I|} \cos \theta \quad \text{and} \quad \frac{2}{\omega C_s} = \frac{|V|}{|I|} \sin \theta$$

where θ is the phase lag, $|V|$ is the rf voltage amplitude, and $|I|$ is the rf current amplitude. In addition, the ratio of the average effective electric field to pressure E_e/p, the sheath voltage V_s, and the electron concentration n_e can be calculated as

$$\frac{E_e}{p} = \frac{|I|R}{\sqrt{2}\ dp}, \quad V_s = \frac{|I|}{\omega C_s}, \quad \text{and} \quad n_e = \frac{d}{Ae\mu_e R}$$

where d is the electrode spacing, A is the electrode area, and e is the charge of an electron.

The above impedance model makes a number of simplifying approximations which are not physically correct. First, the electron concentration is not uniform but varies approximately as shown in Figure 3. The electric field at the interface between the plasma and the sheath increases causing a locally higher E_0/p than the E_e/p which the model calculates which represents only an average value, however, it should be qualitatively correct. Due to the higher E_0/p in the sheath, the higher energy processes are favored, e.g. ionization. Typically, the electron concentration is greatest at the center of the discharge since electrons are produced throughout the dsicharge and gas-phase electron loss processes are often much less than electron loss by diffusion to the electrode surfaces. The power dissipated by ion bombardment of the surfaces at these frequencies (4 to 12 MHz) is relatively minor (less than 3%), as can be shown by the product of the ion flux and sheath voltage (which were experimentally measured and are discussed below). However, the model assumes a constant sheath capacitance and does not include the motion of the ions and electrons within the plasma sheath which leads to nonlinearities that can be readily observed at the lower frequencies. The model does a reasonable job of identifying the trends in electron density, electron energy, sheath voltage, and sheath capacitance which are necessary to the under-standing of the plasma kinetics.

Polysilcon Etching in Cl_2 Discharges

The etching of heavily P-doped polysilicon films has been studied as a function of frequency in a single-wafer, parallel plate etcher with anodized aluminum electrodes. The etching rate was measured as a function of frequency, Figure 4, while maintaining a constant pressure of 0.6 torr, an electrode spacing of 2.5 cm, a Cl_2 flow rate of 5 sccm, and a discharge power of 0.8 W/cm^2. Similar trends were also observed at 0.3 torr. Below about 5 MHz, the etching was anisotropic suggesting that the etching process was controlled by ion-induced processes. Above 5 MHz, the etching was isotropic indicating the dominance of chemical etching, with Cl likely being the primary etchant. The increase in rate above 5 MHz suggests that the fraction of power dissipated in Cl production increases with frequency.

The diagnostic measurements described below were performed in a similar apparatus, but with stainless steel electrodes. The etching rates in this system also increased with frequency above 1 MHz, in constrast to the decreasing rates observed by Bruce(11). Bruce, however, was etching undoped, single crystal silicon while in this study n^+-type polysilicon was used which is expected to be attached by Cl. The amplitude of the rf voltage at the powered electrode is also shown in Figure 4. Using $|V|/2$ to scale the ion energy, the energy should significantly decrease with increasing frequency. In addition, at low frequencies the ions are able to pass through the sheath in less than one rf cycle, thereby increasing the maximum ion enegry. The decreasing ion energy with frequency should reduce the ion-enhanced etching rate which could partially explain the transition from anisotropic to isotropic etching.

The optical emission for peaks associated with Cl_2 (255 nm), Cl_2^+ (386.3 nm), and Cl (837.6 nm) are shown as a function of frequency in Figure 5 where the emission has been normalized for all the peaks at 100 KHz. The trends follow those of the etching rate. Since the total pressure is held constant, the increasing Cl_2 emission suggests that the electron concentration within the plasma increases with frequency or that the electron energy changes causing greater emission. The decline in the Cl emission relative to that of Cl_2 does not necessarily indicate the reduction in Cl concentration since the alteration of the electron energy distribution and the difference in excitation thresholds could dominate.

The positive ion flux to the wafer electrode, shown in Figure 6 as a function of frequency, was determined using an electrometer which measured the ion current passing through a 50 μm orifice in the wafer electrode. Since the diameter of the orifice is less than that of the mean-free-path of the particles within the plasma, the flow is collisionless and equivalent to that impinging upon the wafer. At lower frequencies, the observed flux correlates well with the observed etching rates suggesting that the rate is controlled by ion bombardment. However, in the high frequency region, the flux does not correlate with the etching rate. The large variation of the

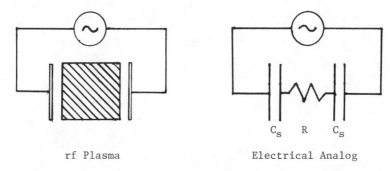

rf Plasma   Electrical Analog

Figure 2. Electrical analog for capacitively powered plasma.

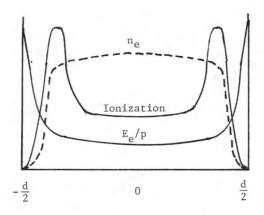

Intra-electrode Position

Figure 3. Expected non-uniformity within a plasma as a function of position between electrodes.

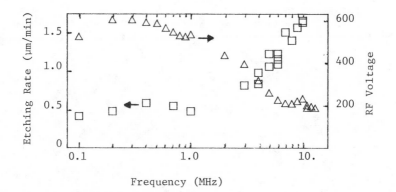

Frequency (MHz)

Figure 4. Etching rate of P-doped polysilicon and rf voltage in a 0.6 torr Cl_2 discharge.

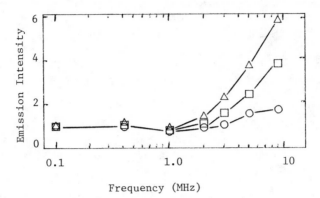

Figure 5. Optical emission from Cl_2 discharge as a function of frequency. Key: □, Cl_2; △, Cl_2^+; and ○, Cl.

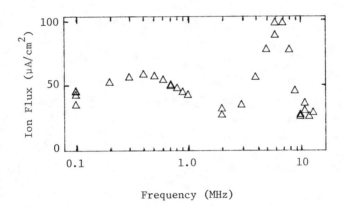

Figure 6. Ion flux to the wafer electrode as a function of frequency.

ion flux is difficult to interpret with the above data alone. Argon discharges measured over the same frequency range indicated a monotonic increase in ion flux reducing the possibility of experimental corruption caused by fields at the orifice or effects associated with the frequency on the ion sheath transport.

Using the impedance modeling of the plasma discussed above, E_0/p and n_e have been calculated using a μ_e of $6x10^4$ torr-cm^2/V-s, Figure 7. As suggested by the increasing Cl_2 optical emission, the electron density rises significantly in the high frequency region, while the E_e/p is found to decrease. Therefore, as the frequency increases, the plasma goes from a lower density of higher energy electrons to a higher density of lower energy electons. In all cases, however, the plasma dissipates the same energy. The higher energy electrons favor processes which have larger threshold energies such as ionization processes, while the lower energies favor the lower energy processes. The direct ionization of Cl_2 to Cl_2^+ requires electron energies of greater than 13.2 eV, while dissociation (producing 2Cl) and dissociative attachment (producing Cl and Cl$^-$) have threshold energies of less than 3 eV. Thus, the dissociative processes which produce Cl are favored at the higher frequencies, while the ionizing processes are likely the primary mechanism by which energy is dissipated at lower frequencies. Comparison of the above conclusions with more quantitative measurements made on O_2 discharges, which undergo similar processes, strongly supports these conclusions(3).

Figure 8 shows V_s, $|V|/2$, and C_s as a function of rf frequency. The difference between V_s and $|V|/2$ are a result of the resistance within the plasma, which declines as a function of frequency. It can be seen that C_s and V_s are complex functions of frequency. Assuming that the sheath width is primarily determined by positive ion transport and that the electron concentration is low within the sheath, Γ_+ at the cathode of a DC discharge is related to the sheath width x by

$$\Gamma_+ = \frac{4\varepsilon_0}{9e}\sqrt{\frac{2e}{m_+}}\frac{V_s^{3/2}}{x^2} \quad or \quad = \frac{9\varepsilon_0\mu_+}{8e}\frac{V_s^2}{x^3}$$

where ε_0 is the permittivity of vacuum. Space charge and mobility limited transport is assumed, respectively(4). Assuming the capacitance is inversely related to x and the average sheath width in a rf discharge has a functionality similar to that of a DC cathode sheath, C_s is related to V_s to a power between -0.75 and -1.5. In a Cl_2 discharge, the Cs and Vs vary inversely as predicted for a constant flux, but the flux varies widely as shown in Figure 6. The Γ_+, as discussed previously, is ideally proportional to the product of n_e and E_e/p, which can be calculated from Figure 7 and rises continuously. The contradictory nature of Γ_+ suggests that the sheath kinetics are more complex than this model, which assumes the diffusion ions to the sheath and a sheath width determined by positive ion transport.

The failure of the above model to predict correct trends suggests fundamental omissions. The discrepancies in the Γ_+ and C_s could be a result of the production of significant concentrations of Cl$^-$ within the discharge above 7 MHz where dissociative attachment should be significant. The presence of the more massive negative charge carrier reduces the diffusivity of the positive ions and removes the constraint of equal electron and ion diffusivities since

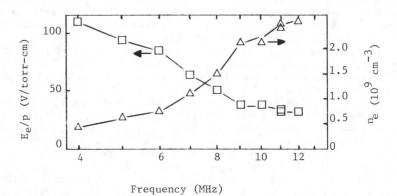

Figure 7. Average effective electric field and electron density calculated from plasma impedance measurements.

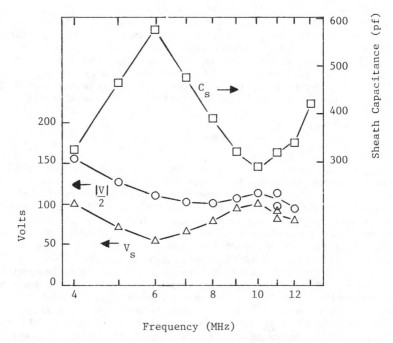

Figure 8. Sheath voltage and capacitance as a function of frequency.

the constraint of net neutrality now involves three species. Assuming that the fraction of negative charge carriers that are electrons is a constant β and that the diffusivity of positive and negative ions is approximately equal, the positive ion diffusivity is

$$D_+^* \approx D_+ \left[1 + \frac{\beta T_e}{T_+} \right] \approx D_+ \frac{\beta T_e}{T_+}$$

The presence of the negative ions reduces the positive ion diffusivity. The presence of negative ions near the plasma sheath would reduce the diffusive flux of positive ions to the sheath.

Similar changes in electron densities and energies have also been observed with variations of pressure at a fixed frequency. The occurance of similar discontinuities in the ion flux suggests the onset of significant Cl^- production. The changes of states observed can be accompanied by hysteresis and instabilities within the discharge which are characteristic of kinetic systems with multiple steady-states.

Chloro-fluorocarbon Discharges. Chloro-fluorocarbons have dissociation mechanisms similar to Cl_2 at similar or lower threshold energies. Preliminary studies confirm that similar changes in n_e and E_e/p can occur. The instabilties observed for Cl_2 etching and chloro-fluorocarbon systems may also be a result of the low energy mechanism for dissociation and/or the production of Cl^-. Fluorocarbons such as CF_4 can undergo dissociative attachment, but typically have a threshold of 5-6 eV, significantly higher than that of the Cl-bearing gases. Therefore, the free radical production and the sheath kinetics should be quite different for these gases, resulting in fundamental differences in the etching behavior.

Acknowledgment

This work was supported by the Semiconductor Research Corporation contract # 83-01-033.

Literature Cited

1.   Greene, J. E.; Barnett, S. A. J. Vac. Sci. Technol. A, 1983, 21, 285.
2.   Coburn, J. W. Plasma Chem. Plasma Process. 1982, 2, 1.
3.   Bell, A. T. In "Techniques and Applications of Plasma Chemistry"; Hollahan, J. R.; Bell, A. T., Eds.; Wiley: New York, 1974, chap. 1.
4.   Chapman, B. "Glow Dicharge Processes";Wiley: New York, 1980.
5.   Winters, H. F.; Coburn, J. W.; Chuang, T. J. J. Vac. Sci. Technol. B, 1983, 2, 469.
6.   Gerlach-Meyer, U. Surf. Sci. 1981, 103, 524.
7.   McFeely, F. R.; private communication.
8.   Reinberg, A. R. In "VLSI Electronics Microstructure Science"; Einspruch, N.G., Ed.; Academic: New York, 1981, Vol. 2, chap. 1.
9.   Cherrington, B. E. Plasma Chem. Plasma Process. 1982, 2, 113.
10. van Roosmallen, A. J. Appl. Phys. Lett., 1983, 42, 416.
11. Bruce, R. H. Solid State Technol. 1981, 24(10), 64.

RECEIVED December 26, 1984

Applications of Oxides and Nitrides of Germanium for Semiconductor Devices

O. J. Gregory[1] and E. E. Crisman[2]

[1]Department of Chemical Engineering, University of Rhode Island, Kingston, RI 02881
[2]Division of Engineering, Brown University, Providence, RI 02912

The reactions of germanium with oxygen and with nitrogen are reviewed. Particular emphasis has been placed on the relationship between the structure and electronic properties of the oxide and nitride films formed on single crystal germanium surfaces under different processing conditions. A summary of the electronic properties reported is presented and some conclusions are offered based on the recent literature.

1. Introduction

Although germanium and its compounds have been studied for more than a century, the reactions of that element with oxygen and nitrogen received particular scrutiny in the early 1930's and again through the 1950's as enthusiasm surged for transistor based electronic circuitry. Once planar technology was developed for silicon, interest in germanium devices, based mostly on alloy junction technology, declined except for a few special applications: particularly in near infrared detectors and low forward voltage rectifying diodes, both of which continue to be widely used and studied (1,2). However, a continuous research effort was maintained through the 1960's and 1970's resulting in a steady, albeit low volume, stream of published information with a significant fraction generated by Russian, Hungarian and Japanese researchers. Examples of the type of information from outside the U.S.A. can be found in references (3,4,5). Much of that information deals with the reaction products of germanium with oxygen or nitrogen as a function of temperature and to a lesser extent pressure. Now that silicon VLSI technology is approaching theoretical limits associated with bulk properties of highly perfect single crystal material, renewed interest in germanium has developed and so the previous research is assuming new importance.

Planar processing technology derives its success from the ability to grow high quality, passivating insulators on the semiconductor surface. This is particularly true in the case of silicon technology, where the thermal oxidation of silicon in oxygen and/or steam at 1000°-1200°C is used almost exclusively to

0097-6156/85/0290-0178$11.50/0

produce a nearly perfect passivating insulator on the silicon surface (6). However, as silicon device elements become smaller and more densely packed, alternative semiconductors such as GaAs and Ge are being considered to make high speed digital and microwave devices. Certain semiconductor properties of germanium, for example, warrant its consideration in complimentary pairs (CMOS and HCMOS) device applications provided that low surface state densities can be achieved at the insulator-semiconductor interface. At room temperature, both electron and hole mobilities are greater in germanium than in silicon. Also, the electron and hole mobilities are more nearly equivalent in germanium, making it an attractive candidate for complimentary device structures, particularly at reduced temperatures (7) where the mobilities increase more rapidly with temperature decrease in germanium. However, before integrated circuits based partially or totally on germanium can be successfully fabricated, a reliable means of forming stable, electrically insulating, passivating layers on germanium surfaces has to be demonstrated. In this paper, the structure, kinetics and formation of "native" oxide and nitride layers on germanium will be reviewed for a variety of deposition methods. The electronic properties of the resulting insulator-semiconductor interface are also summarized as part of this review. While details of all the investigations related to "oxides and nitrides of germanium" are beyond the scope of this review, we have tried to include as diverse a cross-section as possible and to point out where conflicting results exist in the literature. We include the following index of topic headings to facilitate the use of this manuscript as a future reference for the reader.

2. Germanium Oxide Insulators

2.1 Background. At least four methods of forming native oxides on single crystal germanium surfaces have been extensively investigated with the intention of forming passivation layers for germanium based device applications. These are: 1) thermal oxidation, 2) thermal oxidation through a "non-native" capping film (viz. SiO_2), 3) "wet" chemical oxidation, and 4) ion implantation. We will discuss these methods individually after a few remarks about the preparation of "clean" starting surfaces and the oxides that form there-on during post-etching exposure to room air ambients.

When an atomically clean, single crystal surface of germanium is exposed to an oxygen bearing atmosphere at room temperature, rapid uptake of oxygen occurs until about 0.7 atomic layers are formed. Thereafter, the absorption kinetics follow a logarithmic oxygen uptake until a ratio of about 2.5 oxygen atoms per surface germanium atom is reached (8,13). To that point, at least, the layer appears to be highly stable. After the first few atomic layers are formed, oxygen uptake falls rapidly below that of a logarithmic rate equation (8,9,10). To date, the best estimate of the first atomic layer or so indicates two germanium oxygen intramolecular bonds of the monoxide type with the two remaining bonds available for covalent attachment of the GeO molecule to the surface (11,12,13,14). This conclusion, however, is in apparent conflict with the results of von Wienskowski and Froitzheim (15), who indicated O_2 quasi-molecular oxygen bonding to the surface. On uniform surfaces, the monoxide appears to provide uniform coverage (19). After the first few atomic layers, the continuation of oxidation at room temperature results in increased formation of a GeO_2 phase until 10-50 A are formed depending on the cleaning procedure. This is usually achieved in under 30 minutes at pressures at or less than one atmosphere as measured by Green and Kafalas (20). They showed by differential pressure measurements in an experiment sensitive to about $5x10^{14}$ atoms/cm^2, that no oxygen uptake occurred after 28 d of exposure. This implied that surface absorption on the specimen reached saturation during the 30 m which elapsed between etching and sealing in the system. Others who have studied "clean" surfaces exposed to known pressures of O_2 and O_3 have deduced sticking coefficients of 0.03 (14,21) and 0.5 (22), respectively, indicating that the amount of atomic or ionic oxygen present is a factor in the initial formation rate. The presence of water molecules in the ambient is also know to strongly affect the formation and reaction rates of the various oxides but the exact role is as yet undetermined (20,23,25). While the nature of this interface will undoubtedly assume major importance as the technology of germanium advances, we have considered in this review primarily the thermodynamics and kinetics for thicker films ($\geq$ 100 A) and the phase equilibria in the germanium-oxygen system.

There is ample evidence in the literature that the semiconductor/oxide interface, when properly formed, will be well passivated (24,25,26,27) in an electronic sense and, therefore, potentially suitable for germanium based MOS technology. As mentioned above, Green and Kafalas (20) showed that freshly CP4 etched single crystal surfaces stabilized in less than 30 m in

either wet or dry oxygen ambients. Archer (28) observed that
oxidation still proceeded after 28 h but his surface preparation
differed significantly from that of (20). In both cases the
equilibrium thickness appears to fall within the 10-50 A range
usually observed for "natural" oxides. We, also, have monitored
the buildup of "natural" oxides for CP4 and HF-acid etched
surfaces, each followed by DI water rinse and zero grade N_2 blow-
dry. The P-type specimens were of (111) orientation with less than
$5x10^3$ etch pit density. We used ellipsometry to monitor thickness
buildup and index of refraction over a 24 h period. In both cases
a final thickness was observed to be on the order of 30 A and the
index of refraction 2.2. But the HF-etched specimen reached
equilibrium in less than 1 m whereas the CP4 etched specimen
reached thickness equilibrium in approximately 10 m and the index
of refraction continued to change from 1.56 to 2.21 over 30 m The
orientation dependence, if any, of this "natural" oxidation has not
been well established as far as we have been able to determine.
Whatever the conditions of formation, the interface properties of
post CP4 etched oxides are easily modified by successive treatments
in wet and dry oxygen bearing atmospheres (25) or by heating to
high temperature in high vacuum and re-exposing to oxygen ambients
(29).
    Reconstruction of nearly perfect "clean" (100), (110) and
(111) germanium surfaces was interpreted (12,30) to have doubled
the lattice point periodicity in one of the two major directions
while leaving the other unchanged, relative to the undistorted
lattice. This (2x1) surface configuration has also been observed
by others in more recent experiments (13,15). Whether a "clean"
germanium surface is possible, or even desirable, as a step in IC
processing is an arguable point at this time. It is likely that
such a specimen surface would be highly unstable and susceptable to
contamination from local ambient gases. Probably the most
adaptable cleaning cycle at present would be the heating of a
freshly HF-etched and nitrogen dried specimen in flowing forming
gas < to < 550°C T 650°C which would sublime and reduce the
"natural" oxides but still be below the activation threshold for
nitride formation. Hydrogen is not apparently absorbed to any
extent on major crystal planes of germanium or, if it is, would
react locally with surface oxygen to form H_2O thus promoting the
reduction process. The flowing gas is necessary to prevent
deposition of condensed GeO_2 solid on the crystal surface from the
gaseous phase. From work function measurements of "clean" surfaces
the minimum number of surface states that might exist is estimated
to be $5x10^{13}$ (11,32). Even poor passivation of germanium has
reduced this to $1x10^{12}/cm^2$ and some have done considerably better
(33,34,35) using native oxides. Barnes and Banbury, however, have
reported $5x10^{11}$ on vacuum cleaved surfaces (31). The evidence
suggests that the films spontaneously formed on freshly polished
surfaces of germanium crystals (57) are of the tetragonal phase.
Sladkova has reported the index of the 30-50 A surface films,
formed within 20 m of polishing, to be 2.0-2.2. Such films are
insoluble in water and, more significantly, insoluble in HF-acid.
These two observations are consistent with the reported
characteristics tetragonal - rather than hexagonal - or amorphous
GeO_2, of $GeO_2(T)$ rather than $GeO_2(H,A)$, hydrated GeO or $Ge(OH)_2$.

The conclusion is also consistent with the observations of Frantsuzov and Makrushin (58), who showed that these "natural" oxides could not be removed by heating to 925°C specimens which had been cleaned by ion bombardment, annealed and exposed to 10^{-7}-10^{-6} torr of oxygen. No GeO desorption was observed if the specimens were not ion bombarded before the oxygen exposure step. This indicated a highly stable film was present after the chemical cleaning and polishing of the germanium. The transition temperature for tetragonal-hexagonal conversion has been measured to be 1033°C (59) which is significantly higher than the maximum 925°C used to effect "flash" desorption in those experiments.

Generally, it can be stated that surfaces freshly etched in non-oxidizing solutions have a 10-50 A oxide that form immediately with a residual 10^{12} "fast" interface states and likely at least one order more of slow states (8). On the other hand, oxidizing etches tend to leave behind thicker layers and/or remove germanium into the solution*. The experiments of Many and Gerlich (36) have shown that the as-etched surface cannot be characterized uniquely by a single set of parameters since its structure depends on the preparation history of the surface. This explains the differing results of many investigators (37,39). The point may become less important now that several researchers have reported reduced interface state densities and fixed interfaced charges. Electron microscope studies of the surface morphologies of oxides resulting from various chemical preparations has been done by Giber et al. (40).

2.2 Fundamental Germanium-Oxygen Reactions

2.2.1 Germanium Monoxide.   Probably the first observation of a monoxide phase of Ge-O was recorded by Winkler in 1886 (41). Notes on the preparation of precipitated hydrous and anhydrous GeO powders can be found in the study of Dennis and Hulse 42 and are also summarized by Johnson (43).

The gaseous reaction product of germanium and oxygen which occurs at and above 550°C is generally assumed to be a monoxide phase (44) and is responsible for vapor transport and deposition of elemental germanium at cooler regions of the reaction chamber. The temperature where such decomposition deposits are formed has not been reported but is in the neighborhood of 450°C, based on measurements made in our laboratory. The monoxides that are formed chemically in the vicinity of room temperature appear to be readily soluble in HCl or HF but much less so in water (42). Weakly oxidizing solutions have been demonstrated to give surface films on single crystal germanium which are sometimes reported as monoxides and sometimes as a hydrated Ge(OH)$_2$ (8). Since such films are reported to be insoluble in water and HF solutions, it is likely that they are neither of these but rather a stable phase of GeO$_2$(T) as will be discussed below.

The monoxides of germanium are much more ubiquitous than the three dioxide modifications. Some confusion exists as to the

---

* The etch designated CP4, by the way, has become a rather loosely defined term for almost any ratio of hydrofluoric, nitric and acetic acids.

details of the monoxide thermodynamics. While it is generally agreed that GeO is thermodynamically unstable as a solid phase (45), it can be formed by an alternate method at room temperature (43) and its dissociation kinetics are such that it appears metastable at low temperatures. The oxide which forms immediately at room temperature on chemically cleaned germanium surfaces can readily be sublimed at and above 550°C and, in fact, continuous sublimation (and etching) of a germanium crystal surface is observed if small partial pressures of oxygen are included in the ambient atmosphere above that temperature. The formation of thermal etch pits during monoxide formation indicates an orientational dependence on sublimation kinetics.

A solid compound with 1:1, Ge:O can be formed at room temperature by wet chemistry but that compound is generally amorphous and unstable (42). Attempts to make a thermodynamically stable monoxide have been unsuccessful at temperatures to 1400°C (46,47). No well defined x-ray diffraction pattern has been established for the metastable GeO either. Calculations indicate that a metastable monoxide is energetically favored at room temperature (13) and so the lack of stability of GeO formed by wet chemistry might be traceable to water of hydration. Dennis and Hulse have indicated that the stability of the monoxide is considerably improved when wet-chemistry formed GeO is heated in dry nitrogen to 650°C (42). The heat of formation for the monoxide according to Reaction 1 below has been measured to be 55.1 kcal/mole (47). Best evidence to date indicates that thermal reaction got monoxide-formation rates are independent of crystal orientation.

An appreciable vapor pressure ($\sim 10^{-3}$mmHg) of GeO(g) over Ge(s) and GeO_2(s) is established during the oxidation of germanium at 550°C according to the reaction (47):

$$Ge(s) + GeO_2(s) = 2GeO(g) \tag{1}$$

which suggests that thermal oxidation is pressure dependent. The earliest studies of this pressure dependence on oxidation were those of Bernstein and Cubiciotti (44) who suggested that the rate of evaporation of germanium monoxide can be lowered by increasing the oxygen pressure. In the temperature range 575-700°C, they observed that the volatile monoxide could be further oxidized to yield germanium dioxide according to one of the following:

$$GeO(s) + 1/2O_2 = GeO_2(s) \tag{2}$$

$$GeO(g) + 1/2O_2 = GeO_2(s) \tag{3}$$

for oxygen pressures in the range 20 to 400 mm Hg.

2.2.2 Germanium Dioxide. The chemical reaction of oxygen with germanium is similar to the thermal oxidation of silicon which results in the growth of a thin layer of silicon dioxide. As noted above, germanium oxidation is complicated by the formation of an intermediate reaction product, germanium monoxide, which, unlike silicon monoxide, is thermodynamically unstable as a solid and sublimes near its formation temperature (46,47). Therefore, instead of producing a dense protective oxide layer on the surface, a volatile reaction product is formed at the germanium-germanium dioxide interface for oxidation temperatures above 550°C (48). Consequently, no closed native oxide layer is formed on germanium at these temperatures and pressures. The variation between the thermal oxidation of silicon and germanium is as follows:

$$2 \ Si(s) + O_2(g) = 2 \ SiO(s) \tag{4}$$

$$2 \ Ge(s) + O_2(g) = 2 \ GeO(g) \tag{5}$$

and

$$2 \ SiO(s) + O_2(g) = 2 \ SiO_2(s) \tag{6}$$

$$2 \ GeO(g) + O_2(g) = 2 \ GeO_2(s) \tag{7}$$

Because of Reaction 5, Reaction 7 can occur elsewhere in the system rather than just on the crystal or within the solid oxide forming on the crystal surface. This is the major difference between the two systems and results in dichotomy between the reaction kinetics. In the presence of oxygen bearing ambients, Reactions 5 and 1 can proceed to completion at and above 550°C. This is not a thermodynamically stable reaction (47) and is rate limited primarily by the GeO diffusion away from the solid-gas interface. The reaction will continue until one of the components is completely consumed. In an oxidation study by Law and Meigs (48), the effects of pressure and temperature on the oxidation of germanium were expanded to include the effect of crystallographic orientation. Above 550°C, no dependence on crystal orientation was found. However, below 550°C, the (110) faces underwent appreciably more oxidation compared to the (100) and (111) faces. Based on this they concluded that the removal of Ge ions from the lattice and the subsequent transport of these ions through the oxide, controls the oxidation process. The physical interpretation of the orientation dependance is that Ge ions are more readily removed from the (110) faces because of weaker bonding to near neighbor atoms in the underlying planes.

When acceleration of the oxidation process is promoted by increasing the temperature, mixed phase films are obtained: the exact ratio being determined by the particular conditions (viz, partial O_2 pressure, relative humidity, maximum temperature, etc.). Oxidation kinetics have been demonstrated to be independent of dopant type and species (9) but dependent on orientation (60,61) and thus on crystallographic imperfections on the surface of the germanium crystal (21).

2.3 Structure of Germanium Oxides. Similar to the nitride phases of germanium (see below), most phases of germania are isostructural with silica. Oxides of silicon exist in five different condensed phases (62) at one atmosphere pressure. Other high pressure phases have also been identified and have been well studied, mostly by geologists (38), but have not been explored for device processing. For silicon, the relatively low temperature and pressure forming amorphous phase has proved valuable as a processing "tool". Since

the one component system, SiO_2, exhibits considerable polymorphism
(62), it is not surprising that GeO_2 which is nearly isostructural
with SiO_2, also exhibits polymorphism (59). Germania exists in at
least four well established phases: one for a valance state of II,
GeO, and three for a valance state of IV, GeO_2; commonly referred
to as amorphous (A), hexagonal (H) α-quartz and tetragonal (T)
rutile in the literature (45). Both GeO_2 (A) and GeO_2 (H) are
chemically active and readily dissolve in water at 25°C. For this
reason, thermally grown GeO_2 has not been considered useful in Ge
device fabrication to date. The tetragonal modification, first
discovered by Muller and Blank (63), is known to be resistive to
dissolution in both water and HF-acid (59,61,63). There is
evidence that a second monomorph of the GeO_2 (T) phase exists with
a cristobalite structure (50). Laubengayer and Morton (59) while
indicating that a third crystalline phase was improbable, suggested
the possibility of a monomorphic modification. Muller and Blank
(63) postulated that three crystalline polymorphs would be required
to explain their "oxide evaporation" data. From X-ray data, a
cubic GeO_2 (C) phase is also suggested (52).
   The cristobalite-GeO_2 (T) has been obtained by prolonged
heating of amorphous GeO_2 powders at 690°C (51). Some X-ray
evidence for that modification exists (52) and details of the
structure have been worked out by Seifert et al. (53). To our
knowledge no correlations with X-ray diffraction patterns have been
developed with other techniques such as FIR transmission or the
various thin film and surface probes. Measurements of that sort
have always referred to the tetragonal phase determination as the
"rutile" system. Since cristobalite-GeO_2 (T) has been observed
intermixed with α-quartz GeO_2 (H), it is possible that some of the
fine structure of FIR-measurements (54) have been incorrectly
associated with that phase. Further evidence of the incomplete
understanding of GeO_2 (T) has been presented by Stapelbrock and
Evans (55). They showed, by detailed analysis of the u.v.
absorption edge that the band structure of GeO_2 (T) more closely
resembled a SnO_2 (T) structure than the rutile-GeO_2 (T) band
structure calculated by Arlinghaus and Albers (56).
   Laubengayer and Morton (59) determined the relative stability
below 1035°C of the three GeO_2 phases to be
tetragonal → hexagonal → amorphous. More recently, Sarver and
Hummel (64) developed an accompanying GeO_2 phase diagram which
essentially corresponds to that developed by Candidus and Tuomi
(65) and later revised by Trumbore et al. (66)*. Brewer and
Zavitsanos have also contributed to the phase diagram verification
(67). Kotera et al. (68) reported a form of the tetragonal phase
with a rutile structure which was made by the catalyst assisted
transformation of GeO_2(M). Those results were consistent with the
results of Alber et al. (69) in their study of the kinetics of
GeO_2 thin film transformations. They imply that GeO monoxide is an
interim reaction product created by the catalyst that is then
reacted further to form the tetragonal phase. Some interesting
results were presented by Yamaguchi et al. (51) for the
transformation of the amorphous GeO_2 by various thermal and
mechanical treatments.

---

*   That phase diagram is also available in "Constitution of Binary
    Alloys, 1st Supplement" (45).

Conversion of hexagonal to tetragonal (referred to as "soluble" and "insoluble") was reported by Laubengayer and Morton (59) using a hydro-thermal technique in which a hexagonal GeO_2 specimen was sealed in a reaction "bomb" with an equivalent weight of water and heated to 350°C for several days. After a quick quench to room temperature, it was found that in excess of 95% was converted to the tetragonal phase. This was, however, done on powder specimens rather than single crystal germanium and may or may not be related to the technique of citation 5 in references (69), which reports on the "hydrothermal" formation of tetragonal GeO_2 "by heating to temperatures in the 300°-600°C range under several hundred atmospheres of pressure". Solid-solid hexagonal to tetragonal phase transformations are effected by heating GeO_2 (H) in the presence of a mineralizer in the 800-1000°C range at atmospheric pressure (59,64,68). While the intricate chemical reactions occurring in such technique have a great deal of interest to physical chemists the procedures have not historically been useful for device fabrication because of the impurities they introduce. Interested readers are referred to the papers of Alber et al. (69) and Laubengayer and Morton (59) as a starting point.

2.3.1 Oxide Identification. The X-ray diffraction patterns for the various crystalline forms of GeO_2 can be found in ASTM card files (52). As with SiO, no crystalline structure has been observed for the GeO monoxide phase.

The major phase variations of Ge-O can readily be identified by far infrared (FIR) transmission measurements in the 8 to 40 micron range (54,70,71). Hexagonal and amorphous phases have similar absorption minima with the crystalline modification showing fine structure definition, as well as narrowing of the major absorption minimum. Tetragonal GeO_2 does not appear as strongly defined, possibly because of intermixing with amorphous phase, mixtures of two monomorphs, or the inclusion of impurities during the preparation of specimens used for FIR (usually done by chemical precipitation from solution containing dissolved dioxide). An excellent superposition of amorphous, hexagonal and tetragonal FIR transmission curves is given in (71) and finer details are available from (54). In Figure 1 we have reproduced our own measurements of FIR transmission and include measurements on commercially prepared powder monoxide (72). The commercial monoxide powder shows evidence of some hexagonal or amorphous dioxide as well. Identification of the molecular vibrations responsible for some of the structure in the FIR traces has been provided by Lippencott et al. (54). Differential FIR has been successfully employed by us and others to obtain identification of oxide films as thin as 200 A on the surfaces of 500 micron thick Ge specimens, provided the germanium itself is sufficiently free of surface damage and/or free carrier absorption.

For thinner films, index of refraction measurements can be used to distinguish between amorphous (1.60) hexagonal (1.71) and tetragonal (2.05) phases provided the films are of single phase composition. This condition is unfortunately not always satisfied for very thin films as noted earlier. Determination of the monoxide phase in thin films is more difficult. We have been

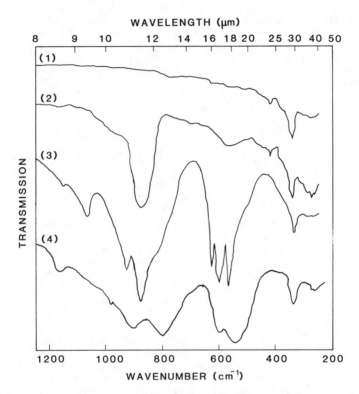

Figure 1.   Infrared transmission spectra for: (1) Bare single crystal germanium; (2) Single crystal germanium with amorphous-GeO_2; (3) Powdered hexagonal-GeO_2; and (4) powdered GeO.

unable to identify a reliable index of refraction for GeO, although a value of 1.78 has been reported in the literature (73).

The X-ray diffraction patterns for GeO_2 (H) and GeO_2 (T) have been accurately determined and can be used to identify films as thin as 500 A (using low angle diffraction). Here again, the monoxide phase cannot be identified by this means since no diffraction pattern has been observed for that phase. Some other non-destructive techniques have been used such as low energy electron diffraction (LEED), electron loss spectroscopy (ELS), Raman scattering, etc. but usually they are so sensitive to contamination that the results cannot easily be used for simple phase identification. Such techniques are therefore more useful for physical property studies.

2.4 Oxidation Kinetics. The relationship of temperature and pressure to the oxidation kinetics of single crystal germanium surfaces was intitally explored in detail by Bernstein and Cubicciotti in the temperature range 575-700°C for O_2 partial pressures of less than one atmsophere (44). They determined that the kinetics of the Ge-O reaction in that range did not conform to any of the known rate laws for metal oxidation. Law and Meigs extended that study in the 450-700°C range to include the three lowest index plane surfaces (48,49). The summation of results from those two reports established that 1) there is little if any oxygen uptake at 450°C; 2) essentially identical rates above 550°C for (100), (110) and (111) surfaces and 3) only oxidation of (110) at 550°C. Below 550°C the oxidation rate was controlled by two factors: removal rate of germanium atoms from the lattice and diffusion through the solid oxide forming on the surface. (Both elemental Ge and GeO can diffuse outward at the same time that O_2 is diffusing inward.) Orientation dependence of the removal rate of germanium was rather subtlely, but convincingly, inferred from the 550°C observations and simplified models of the surface atom bonding for the three major orientations. On (111) and (100) faces, each surface atom is bound to three and two atoms, respectively, in the next layer. On the (110) surface only one bond occurs to the subsurface layer and intra-surface plane bonding is strong. Thus, removal of individual surface atoms weakens the bonding of the adjacent surface atoms as well. Above 550°C the orientation independent oxidation rates were an inverse function of O_2 partial pressure and could most readily be explained by the rate of diffusion of the gaseous monoxide, GeO, away from the solid/gas interface. That conclusion was verified by using mixtures of O_2 and N_2 for the ambient and demonstrating that total pressure affected the rates regardless of large variations in oxygen partial pressures. It must be noted that those two investigations, (48,49), included the oxygen uptake rates of the O_2 that was involved in the GeO (g) transport of germanium to other parts of the system. The rate equations therein developed do not represent the true formation rate of solid oxide films on the surfaces of the solid germanium.

Sladkova using optical measurements on the surface films per se, established that a linear rate existed at 525°C and a parabolic rate at 550°C and above (57,60) confirming again that the GeO sublimation and gaseous phase diffusion was a rate limiting factor.

Sladkova also found substantial variation in index of refraction values of the "naturally occurring oxide" (1.69–2.05) for rather small changes in probe wavelength (5461 A and 5893 A) and it is tempting to speculate that the hexagonal and tetragonal phases of GeO_2 were independently observed particularly in view of the remarkable coincidence with the measured indices for those two phases individually prepared (see Table I).

Table I. Properties of the Various Germanium Oxide Phases

Parameter	GeO	Hexagonal $GeO_2(H)$	Tetragonal $GeO_2(T)$	Amorphous $GeO_2(A)$
Structure	Amorphous	Hexagonal	Tetragonal	Amorphous
Crystal type	na	Low-Quartz	Rutile*	na
Molecular Weight	88.59	104.59	104.59	104.59
Density	?	4.23–4.28	6.24–6.26	3.64
Melting point	Sublimes at 550°C	1116±4°C	1086±5°C	supercooled liquid
Lattice Param. a	na	4.987 A	4.395 A	na
c	na	5.652 A	2.860 A	
Index of Refraction	1.78	1.67–1.74	2.05–2.10	1.61

* Evidence exists that a second tetragonal phase also exists with a cristobalite crystal type. No distinction was found in the literature for the dielectric and optical properties of the two monomorphs.

Little work has been done on the oxidation rates of various crystal planes of germanium for oxygen pressure greater than one atmosphere. Crisman et al. (61) have reported orientational dependence on (111) and (100) germanium surfaces using the high pressure oxygen (HPO) technique first reported for silicon oxidation by Zeto et al. (74). Oxygen pressures to 1360 atmospheres were used but only one temperature (550°C) was explored. They, also, were unable to relate their observations to any known rate equation but did point out that the measured results were unrelated to the oxidation rate equation established by Deal and Grove (75) for silicon-oxygen reactions. A peak in the index of refraction curve vs. pressure was also reported by (74) for pressures in the vicinity of 340 atmospheres (5,000 psia). Recent studies at pressures to 1020 atmospheres have shown that significant oxide films are formed at temperatures as low as 400°C

and that the morphology is orientation dependent (76). As will be
discussed in section 2.5 below, interfaces prepared by HPO have
been of "good" electrical quality and have been successfully used
for construction of MOSFET devices. In addition, HPO oxide
surfaces appear to be an appropriate starting point for the
formation of nitride films on germanium (see Section 3.5. and
references (72,80). Electron microscope studies have shown them to
be extremely uniform and featureless at 10,000 X (80).

2.5 Preparation and Properties of GeO2. The various nonthermal
techniques described above are useful in preparing specimens of
predominately known phase for individual study. They are, however,
of marginal value for thin film production on single crystal
surfaces since such reactions are further confused by the unlimited
supply of germanium. No interfaces of good electrical quality have
to date been reported using the non-thermal conversion techniques.
     Although not of primary concern in this review a number of
experimenters have developed information on the oxidation rates of
amorphous and polycrystalline films of germanium. While there is
some disagreement on the role of porosity in the oxidation rate of
such films, (17,18) the kinetics of the reaction appear to be
strongly dependent on the morphology of the films and the ambient
atmospheres to which they are initially exposed. Based on that
information it would appear that implant areas should be annealed
in situ or at least removed from vacuum to a controlled environment
until final surface preparation is affected. This is particularly
true of photovoltaic and photo-conductive devices where the
uniformity of oxide, interface moisture content, uptake of carbon
complexes etc. strongly affect the surface recombination currents
and hence the device performance (77).

2.5.1 Direct Thermal Oxidation. We have referenced earlier
several investigations which had directly reacted oxygen with
single crystal germanium surfaces for the purpose of developing
oxidation rate information (48,49,60,69,70). Studies of direct
thermal oxidation have also been done to analyze the various
parameters which characterize the electronic nature of both the
insulating (and electrically passivating) film and the film
semiconductor interface. Because of the sublimation of GeO during
thermal oxidation, the early attempts to produce passivating layers
on single crystals by this means were not entirely successful (78).
Ercil (27), using oxygen ambients at high pressures (up to 1360
atmosphere) was able to oxidize both (111) and (100) surfaces and
to demonstrate that interfaces, thus formed, had superior
electronic characteristics to what had been reported previously.
While no interface state densities were calculated for the
specimens, Ercil using the Zerbst transient capacitance technique
(79), demonstrated surface recombination (actually generation)
velocities (s) that were one to three orders of magnitude less than
the 100 cm sec^{-1} usually accepted for damage-free germanium
surfaces with "natural" oxides. In those measurements, the s-
values for p-type were found to be 3 to 4 times greater than the n-
type crystals: a ratio also reported for chemically prepared
surfaces (37). Spectral response measurements of photovoltaic
devices with and without the high pressure oxides were used to

qualitatively verify that surface recombination currents had been
significantly reduced by the presence of the HPO formed layer. The
HPO technique (61) was subsequently used by Crisman (35) to
construct Ge MOSFET's. Interface "fixed" charge densities were
measured to be $1-3x10^{11}cm^{-2}$ in that study and maximum midgap interface density of states were $7x10^{11}cm^{-2}eV^{-1}$ with $2.8x10^{11}cm^{-2}eV^{-1}$
for n-type and $2.3x10^{11}cm^{-2}eV^{-1}$ for p-type being the lowest
observed values. Surface mobilities for the MOSFETs of that study
were measured to be 20% of the bulk for n-type carriers. While the
"fast" state densities were approaching values sufficiently low for
effective device fabrication, there were large variations in the
"slow" state density (from $1x10^{11}$ $cm^{-2}eV^{-1}$ to $5x10^{13}$ $cm^{-2}eV^{-1}$)
which would severely hinder the switching speeds of such devices.
Further work is obviously required to understand the source of the
slow states.

2.5.2 Thermal Oxidation Through SiO$_2$ Cap. Several investigators
have explored the feasibility of forming an oxide of germanium
beneath a pre-deposited silicon dioxide layer (33,34,70). Kuisl
(81) reported results on germanium surfaces which were annealed
between 600°-800°C in oxygen after first capping the surface with a
SiO$_2$ layer produced at 500°C by the reaction of SiH$_4$ and NO$_2$.
Layers of 150 to 250 A were used for the encapsulation. FIR-
transmission after oxygen anneal showed a germanium dioxide
absorption consistent with four fold (hexagonal) coordination.
Using an interference measurement technique (82), a parabolic rate
equation was developed which fit the equation:

$$thickness = a \ (time)^{\frac{1}{2}} + b$$

where "a" is a function of annealing temperature and "b" is a
constant depending on the thickness of the capping layer. This is
similar to the rate equation for Si oxidation (75). It was assumed
that oxygen diffused through the SiO$_2$ layer to oxidize the
germanium below. This may be a rather simplistic conclusion based
on more recent measurements in our laboratory and by Wang and Joshi
(83) which show considerable interdiffusion of Ge and Si into each
others' oxide.

Kuisl (81) did not report depth/composition data for his
silica/germania/germanium structures. He did, however, measure
rather high conductance through the films and distortions in
capacitance-voltage curves which indicated poor surface
passsivation characteristics. The somewhat low values for index of
refraction of the SiO$_2$ surface layer might account for this by
providing localized breakdown in the films due to nonstoichiometry. Wang and Gray (84) extended the studies of (83) to
include comparisons of anneals in Ar, O$_2$ and H$_2$/N$_2$ forming gas.
Interface state densities were shown to decrease from $10^{14}cm^{-2}eV^{-1}$
to $5x10^{11}cm^{-2}eV^{-1}$ for the non-oxygen vs. oxygen anneals. Depth
profiling disclosed significant germanium oxide formation only for
the oxygen annealed specimen although once again Si/Ge-oxide
interdiffusion was evident. Respectable surface mobilities were
reported in that study with values approaching 10% of bulk at room
temperature. Toshitaka et al. (1) have reported reduced interface
state density and dark current using CVD SiO$_2$ encapsulation

followed by an oxygen anneal. The optimum annealling temperature was reported to be in the vicinity of $600^{\circ}C$. Using a similar construction technique, Schroder (85) reported a germanium based linear charge coupled device (CCD) array which was successfully operated at 10-300 KHz clock frequency in the 100-265°K temperature range with charge transfer efficiency in excess of 95%. A surface state density of $10^{11}cm^{-2}eV^{-1}$ was reported by Schroder for his processing. Very detailed studies of SiO_2-GeO_2-Ge MIS structures were presented by Jack et al. (34) using a variation of deep level transient spectroscopy (DLTS). The specimens were prepared similar to the CVD/oxygen anneal process reported by Wang and Gray (84). The specimens were p-type dislocation-free (100) germanium (10-18 ohm-cm) with a 1500 A CVD SiO_2 cap deposited at 450°C and annealled at 600°C for 2 hours. A wealth of information is available in that study but an important point to note is the low value reported for (fast) interface density of states: for specimens constructed by that processing: $5-10x10^{10}cm^{-2}eV^{-1}$ near mid-band to $100x10^{10}cm^{-2}eV^{-1}$ near the valance band edge. The observed "slow" states were measured to be a factor of 5 to 10 higher.

The various results suggest that the SiO_2 used for capping before high temperature oxygen anneals might be a promising solution both to GeO sublimation and hydroscopic problems associated with germanium native oxide passivation. For this system to be useful, better understanding of the chemistry and thermodynamics of the Si/Ge-oxide interdiffusion will be required but the processing is in accord with current technology.

2.5.3 Ion Implantation. Ion implantation of O_2^+ oxygen into germanium has been demonstrated by Stein (86). The FIR transmission curve of that study shows amorphous four-fold coordinated GeO_2 bands at $800cm^{-1}$ and $570cm^{-1}$ with a weaker absorption band evident at $700cm^{-1}$. The weaker band is at the established position for the tetragonal GeO_2 phase (54). The tetragonal absorption was removed by a 500°C anneal for 20 m in air and none of the GeO_2 (H) fine structure was observed before or after the anneal. No information was presented with regard to the effects of water or HF-acid on the implanted layers. The general shape of the FIR transmission trace indicates that considerable Ge-O bonding is effected during the implant even though the temperature was never more than 200°C (as deduced from annealing threshold temperature for post-implant absorption changes). No electrical performance data was presented for interfaces of the implanted oxide/semiconductor structure. Appleton et al. (87) have reported severe damage structure associated with ion implantation and attributed such damage to vacancy conglomoration in the ion implanted region. The conglomorates can contribute to deep surface voids which considerably deteriorate the electrical properties of interfaces or p/n junctions formed by implantation. Extensive annealing of p/n junctions tends to reduce the deleterious effects of implantation in germanium but this is probably due to solid diffusion of impurities away from the damage region rather than removal of the damage per se (16). Such treatment would probably not be effective for field effect devices which depend on rather perfect surfaces for good performance. Implanting at temperatures in the vicinity of liquid nitrogen eliminates the voids and surface pitting that are observed when implanting at higher temperatures. Holland et al. (88) suggested that immobilizing the vacancies by lowering the temperature allows the self-interstitials which are mobile at much lower temperatures, to recombine locally with vacancy sites. In any case, exposure of the implanted germanium surface to residual gas can result in considerable local impurity absorption in surface voids and suggest that care should be exercised until damage annealing is effected (17).

2.5.4 Wet Chemical Oxidation. Formation of the various oxide phases of germanium by wet chemistry has been studied by a number of investigators and these studies have included both the creation of individual phases in the form of powder precipitates and phase conversion on the surfaces of single crystals (46,59,63,64,69). The former of these have been used primarily to demonstrate phase transformations and to produce large quantities of specific phases for oxide properties studies. These will not be discussed here. Single crystal surfaces have been used in several investigations of oxide formation and properties. Non-thermal oxidation of single crystal germanium has been demonstrated in several different oxidizing media. Nitric acid and hydrogen peroxide based solutions have been used as well as anodic electrolytes.

Edelmen et al. (89) has shown that concentrated nitric acid gave hillocks on a (111) germanium surface that had been lapped, polished and then etched in a CP4 type of solution. Such hillocks were similar to those observed on silicon after Syrtl-etching. Hillock densities corresponded roughly to the starting surface etch pit densities and were also observed to be concentrated along surface scratches. While the predominant oxide was hexagonal, the fraction that was tetragonal appeared to have a definite orientation (not reported). The heterogeneous nature was demonstrated by etching in boiling water and HCl. Subsequent dry oxygen treatment at 700-750°C increased the amount of tetragonal but did not improve the uniformity of coverage. In another experiment using nitric acid oxidation of (100) and (111) Ge, Valyocsik (90) obtained somewhat different results from those reported by Edelman et al. He observed, by three different techniques (FIR transmission, X-ray diffraction and low angle election diffraction), that only $GeO_2(H)$ was formed and that the oxides were polycrystalline with complete coverage for solution in the 7.0N to 15.6N HNO_3 range. Film uniformity improved noticably however as the degree of acidity increased. The (100) surface oxidized at lowest normality but approached the same absolute thickness at the higher normalities. No electrical properties for the films or the interfaces were reported in either study.

Glassy hydrated germanium oxide has been reported by Heidenreich to result from H_2O_2-HF-H_2O solutions (91). Using such solution Kiewit (92) explored the effect of HF concentration on growth rate. He and others (93) determined that neither conductivity type nor mechanical damage from polishing significantly affected the growth rate and the effect of HF concentration was only marginal. Smooth films could be produced up to 2000 A thick at which point "cracking" was observed. Re-immersion of thermally dried films in HF-acid resulted in removal of the oxide implying that water of hydration was a significant factor in the initial formation. That conclusion is further supported by the observation that heating to several hundred degrees decreased the thickness. Films were determined by reflection-electron diffraction to be entirely amorphous, presumably both before and after heating. No electrical properties were presented for either the films per se or the interfaces.

Ellis has reported the surface recombination velocity of H_2O_2-HF-H_2O treated surfaces (94) to be on the order of 10^2 cm sec^{-1} for 1500 A thick films formed in etches with reduced oxidant content. The value of "s" was further reduced by exposure of the film to HF acid fumes for about 10 sec. This resulted in 5-10 cm sec^{-1} values which persisted for several hours. That value is reasonably consistent with measurements of Loferski and Rappaport (96) on freshly HF-etched surfaces without initial oxidation films. The low values of s on those films, however, only lasted for tens of seconds once the HF vapor was removed. The large variability of s in these and other experiments (25,27,37,99) indicate that the pre-oxidation treatment is also an important factor in the final interface condition and that the interface chemistry is still only vaguely understood.

Anodic oxidation of germanium has been studied in both acid and alkaline water based solutions (97), but the most successful oxidations on crystalline germanium have used various anhydrous electrolytes with controlled amounts of H_2O added. Orientation and conductivity type independence of oxidation rate was demonstrated on anodic films grown to 1240 A by Zwerdling and Sheff (98). Wales later showed (93) uniform films to 7000 A could be produced which were limited only by the eventual breakdown of the films per se in the anodization field. Those films were described as "found by X-ray analysis to be GeO_2" but precisely which modification was not reported. Presumably they were crystalline in nature rather than amorphous. Story (23) studied the effects of H_2O on anodic oxidation and concluded that water diffuses through the film and reacts electrochemically at or near the Ge-GeO_2 interface to produce GeO_2. Diffusion of water resulted in film growth at the Ge-GeO_2 interface rather than at the GeO_2-solution interface. No determination of structure or composition of the films was presented but it is reasonable to assume that Story's films were similar to those of Wales since the experimental arrangements were almost identical. Wales reported for his "dry" films, dielectric strengths to $2.1x10^6$ V/cm and film resistivities to $7.4x10^{13}$ ohm-cm (95). This is in contrast to $3.5x10^6$ V/cm and $2.3x10^{10}$ ohm-cm for wet anodic films (93,95). A general but not absolute observation from the variety of information on wet chemical oxide formation is that the presence of water, either in the liquid or vapor state, enhances the oxide formation rate but is deleterious to the electrical properties of the films. Some descriptions of the various chemical preparation techniques for anodic oxidation can be found in the book on the subject by Young (97).

A summary of the best values for the electrical interface properties are presented in Table II for specimens made by the different techniques described above. Ion implantation is not included in that summary since, as yet, most of the properties have not been measured for that oxide formation technique.

Table II. Reported Values for Various Properties of Native Oxides
and Interfaces on Germanium

| | Oxidation Technique | | |
	Direct Thermal	Thermal, Through SiO$_2$ Cap	Wet Chemical
Fixed Surface Charge (#/cm^2)	1x10^{11} ($\underline{35}$)	9.8x10^{10} ($\underline{34}$)	NR
Midgap Density of States (#/cm^2-eV)	7x10^{11} ($\underline{35}$)	0.5-10x10^{11} ($\underline{34},\underline{84},\underline{85}$)	NR
Surface Mobility for n carriers at R.T. (% of bulk mobility)	18 ($\underline{35}$)	15 ($\underline{33}$)	NR
Surface Recombination Velocity (cm/sec)	0.6 ($\underline{27}$)	$\sim$ 100 ($\underline{84}$)	30 ($\underline{94}$)
Dielectric constant	4.5 ($\underline{27}$)	5.3 ($\underline{34}$)	6.4 ($\underline{95}$)
Dielectric strength (V/cm)	3.2x10^6 ($\underline{35}$)	NR	3.5x10^6 ($\underline{93}$)
Maximum Total Thickness (A)	5500 ($\underline{27}$)	4000 ($\underline{85}$)	(NHO$_3$) 10000 ($\underline{90}$) (Anodic) 7000 ($\underline{93}$)

NR = not reported

3. Germanium Nitride Insulators

3.1. Background. The fabrication of a higher speed metal-insulator
semiconductor field effect transistor based on germanium, depends on
the formation of a suitable thin gate insulator. Thin layers of
germanium nitride formed (or deposited) on germanium are well suited
for this purpose (as are silicon nitride films formed on silicon
($\underline{100}$)) and have certain advantages over the properties of germanium
dioxide. Germanium nitrides can be prepared by exposing clean
germanium wafers at high temperatures to anhydrous ammonia ($\underline{7},\underline{10}$)
either in a pure state or in ammonia:nitrogen mixtures. Other
techniques include nitriding thick and thin germanium dioxide films
in ammonia ambients ($\underline{72},\underline{102}$), implanting nitrogen ions directly into
the germanium surface followed by a high temperature anneal

(103),depositing non-stoichiometric germanium nitride and germanium oxynitrde films by reactive sputtering of germanium in hydrazine or nitrogen (104,105,106), reacting germanium tetrachloride with ammonia to produce CVD-germanium nitride films (107,108), exposing germanium wafers to flowing hydrazine to produce a vapor deposited germanium nitride (109) and reacting germanium tetrachloride or germane with active nitrogen produced by an electrical discharge (110). Of these the direct thermal nitridation of germanium or germanium dioxide using anhydrous ammonia is of particular interest due to the ease with which this process could be incorporated into conventional integrated circuit fabrication and thus these methods will be reviewed in greater detail. Emphasis will be on nitridation mechanisms, structure and composition of these films and their electrical properties.

3.2. Fundamental Germanium-Nitrogen Reactions. The thermal nitridation of germanium to form germanium nitride films is similar to the reaction between silicon and nitrogen to form silicon nitride, which is considered isomorphic with Ge_3N_4 (111,112). However, the reaction between nitrogen and germanium is somewhat more complicated due to the decomposition of germanium nitride at temperatures well below the melting point of germanium (960°C). Such is not the case for silicon nitride decomposition which occurs at temperatures well above the melting point of silicon (1410°C). Juza (113) has calculated the equilibrium dissociation pressure (or fugacity) of nitrogen over germanium nitride at 600°C from the data of Johnson and Ridgely (114) to be approximately 10^8 atmospheres. From a thermodynamic viewpoint, then, it should be considered impracticable to grow thermal nitrides on germanium at these temperatures. Since this is obviously not the case, there must exist a kinetic barrier to the decomposition reaction which allows these nitrides to be formed on germanium surfaces at temperatures greater than 600°C. Using ammonia as the nitridant, Johnson (115) observed a sluggish nitride forming reaction rate at 650°C and a much more rapid reaction at 700°C. However at 850°C, the nitride rapidly decomposed. With such behavior, it becomes important not only to consider the thermodynamics of nitridation but the kinetics as well.

The preparation of Ge_3N_4 by direct thermal reaction of germanium with NH_3 was reported in 1930 by Johnson (115). Germanium powder and ammonia were reacted at 700°C to form the nitride according to the equation:

$$3Ge + 4NH_3 = Ge_3N_4 + 6H_2 \qquad (8)$$

Johnson also reported that in the absence of hydrogen, the reaction product Ge_3N_4 slowly decomposed at 600-700°C in vacuo over a period of several weeks to produce nitrogen and germanium according to the reaction:

$$Ge_3N_4 = 3\ Ge + 2N_2 \qquad (9)$$

This led to the conclusion that although the nitride is thermodynamically unstable, a kinetic barrier to the dissociation of the nitride is responsible for its existence at these

temperatures.   Morey and Johnson (116,117) in a subsequent
investigation, studied the decomposition of germanium nitride over
a wide range of temperatures and concluded that the rate of
dissociation increases with increasing temperature.   They also
explained the thermal stability of germanium nitride at 500°C in
terms of an exceedingly slow decomposition reaction rate.

The preparation of a less stable nitride phase, germanous
nitride (Ge_3N_2), was first reported by Johnson and Ridgely (114) in
1934 by the decomposition of germanous imide in ammonia.   However,
they were unable to prepare this nitride directly from the elements
and noted that germanic nitride (Ge_3N_4) exhibited superior
properties to those of Ge_3N_2; i.e. Ge_3N_2 reacted readily with water
and was unstable in air at elevated temperatures.   Storr et al.
(110) later prepared the same nitride by vapor deposition involving
gas phase reactants.   They formed Ge_3N_2 by reacting either
germanium tetrachloride ($GeCl_4$) or germane (GeH_4) with atomic
nitrogen produced from an electric discharge.   With this technique,
the deposition of nitride films on germanium is possible but the
properties of Ge_3N_2 do not warrant its use as an insulator for
devices.

More recently, Hua et al. (101) and Igarashi et al. (102) have
explored the possibility of forming germanium nitrides on germanium
by direct thermal reaction with both NH_3/N_2 mixtures and pure NH_3.
Hua et al. found that $NH_3:N_2$ mixtures in the range 1:1 to 2:1 and
growth temperatures between 600 and 650°C provided the best surface
morphology from a device fabrication viewpoint.   They determined
that germanium wafers, pre-etched in H_2O_2 produced largely
amorphous nitride films which were relatively smooth and uniform.
Igarashi et al. on the other hand, reported the growth of partially
crystallized nitrides when germanium was placed in a pure ammonia
ambient at growth temperatures above 700°C.   This suggests that the
specific reaction conditions as well as the kinetics can influence
the surface morphology and crystallinity.   In particular, the gas
composition, temperature, flow rate and wafer preparation influence
the morphology by promoting the nucleation of different nitride
phases the growth of existing phases on the surface.

Nitride films formed on silicon by thermal reaction are
crystallized when the growth temperature is too high or growth time
too long (100,118).   Similar observations of germanium nitride
films on germanium wafers have been made by Hua et al. (101), who
also related substrate preparation to a possible preferential
crystallization of a single nitride phase.   While there is no
information on the epitaxial growth of Ge_3N_4 on Ge, it might be
possible in light of reported epitaxial growth of Si_3N_4 on Si
(118,119,120).

3.3. Structure of Germanium Nitrides.   Thermally grown nitride
films may be amorphous or crystalline, depending on the growth
conditions. Crystalline Ge_3N_4 exists in two hexagonal forms,
α-Ge_3N_4 and β-Ge_3N_4, which are isomorphic with α-Si_3N_4 and β-Si_3N_4
(121,121).   In both the and forms the basic building unit is
the GeN_4 tetrahedon joined together in such a way that each
nitrogen atom is shared by three tetrahedra.   Ruddlesden and Popper
(111) determined that heating germanium in NH_3 at 750°C produced
largely α-Ge_3N_4 and heating GeO_2 under the same conditions produced

largely β-Ge₃N₄. In their complete structural analysis of germanium
and silicon nitrides, they determined that both α and β-Ge₃N₄ have
a phenacite-type structure (P31C space group) and are related in a
manner which is similar to the cristobalite ↔trydimite polymorphic
phase transformation in the system SiO_2. Based on this argument,
they suggested that the α- and β-nitrides are merely low and high-
temperature polymorphs of Ge₃N₄. However, Grieveson et al. (121)
and Wild et al. (122) later reported that the α- and β-forms were
"high oxygen potential" and "low oxygen potential" modifications.
Based on their findings, they proposed that α-Ge₃N₄ is an
oxynitride with oxygen atoms replacing some nitrogen atoms in the
lattice leaving other sites vacant. More recent investigations
(123,124) clearly indicate that, although appreciable amounts of
oxygen can substitute for nitrogen in α-Ge₃N₄, there is no
structural requirement for it. The unit cell dimensions of α- and
β-germanium nitride are presented in Table III. This table shows
that the essential difference between the two forms is that the c-
dimension of the α-form is approximately twice that of the β-form.

Table III. Unit Cell Dimensions for Germanium Nitrides

Modification	a (A)	c (A)
α - Ge₃N₄	8.1960	5.93
	± 0.0008	± 0.0005
β - Ge₃N₄	8.0276	3.0774
	± 0.0005	± 0.0002

Wild et al. (125) coupled X-ray diffraction studies with
infrared transmission studies to investigate structural
modifications of β-Si₃N₄ and β-Ge₃N₄ that might not be detected by
X-ray diffraction alone. A comparison of infrared spectra between
these two nitrides confirmed the nitrides are indeed isomorphic.
They each exhibited six absorption bands of approximately the same
intensity, with the β-Ge₃N₄ bands shifted to lower wave numbers
relative to those of β-Si₃N₄. This shifting was attributed to the
larger mass of germanium and thus the smaller force constant of the
Ge-N bonds compared to the Si-N bonds.
     Infrared transmission measurements have been extensively used
to identify the various phases that can exist in the system Ge-N
and Ge-N-O, particularly in the 8-40 μm range (72,80,126,127,128).
The crystalline nitride phases (α and β) show considerable fine
structure and a narrowing of the major absorption minima when
compared with amorphous nitrides. The infrared absorption spectra
of crystalline Ge₃N₄ prepared by the thermal nitridation of Ge
powder in NH₃ showed that the spectra of the   polymorphs
differed in both number and position of the absorption bands (126).
The spectra show sharp peaks at 730 and 770 cm⁻¹, corresponding to
the main absorption bands of the α- and β-polymorphs, respectively
(126,127). Another notable difference between the spectra is the
absence of a band at 910 cm⁻¹ for the α-phase (126). Both spectra
exhibit a main absorption band near 840 cm⁻¹ which is attributed to

the asymmetric stretch of the Ge-N bond in the GeN_4 tetrahedron. The plethora of low wave number absorption bands in the range 250–500 cm^{-1} are associated with lattice absorptions or deformation oscillations (128) and are usually observed in predominantly crystalline material.

Amorphous germanium nitrides prepared by CVD (107,108) show almost no FIR fine structure and typically exhibit a rather broad absorption band which is centered between 720 and 750 cm^{-1}, depending on deposition temperature. This minimum shifts to smaller wave numbers as the deposition temperature increases, but always remains in the range of wave numbers corresponding to crystalline Ge_3N_4. Absorption spectra reported by Stein (103) for germanium nitride films prepared by ion implanted nitrogen are very similar with the exception that the main absorption band was shifted to a smaller wave number, due to a lower nitrogen content in the implanted layer (or deviations from stoichiometric Ge_3N_4). Stein (103) also determined that the implantation of NO^+ into germanium produces oxynitride layers with an infrared absorption band position between the nitride and oxide absorption bands. That spectrum was different from the spectra, reported by us, for thick GeO_2 layers nitrided in NH_3 ambients (72,80). In those experiments, we observed the development of absorption bands at 775 cm^{-1} and 730 cm^{-1} which did not shift with further ammonia treatment. These peaks correspond to the α,β-polymorphs of Ge_3N_4 and, unlike Stien's implanted layers (103), no intermediate band positions were observed indicating that a nitride as opposed to an oxynitride resulted from the processing. The relative amounts of the nitride phases were a function of temperature, gas composition and flow rate. A typical FIR transmission spectra of such a nitride film and the oxide from which it was formed is shown in Figure 2. Thus, the development of either the α- or β-modification is dictated more by the particular processing parameters than by thermodynamic considerations as discussed below.

3.4. Thermodynamics of Germanium Nitrides. Thermodynamic calculations suggest that germanium nitride is metastable with respect to germanium and nitrogen and thus the nitride cannot be formed directly from the elements at pressures experimentally attainable (115,116,128). The free energy of formation of Ge_3N_4 from the elements according to the equation:

$$3Ge + 2N_2 = Ge_3N_4 \tag{10}$$

can be deduced by combining the experimentally determined free energy of formation for Reaction 8 with the free energy of formation of ammonia for Reaction 11

$$N_2 + 3H_2 = 2NH_3 \tag{11}$$

which has the form

$$\Delta G = -26,000 + 56.2T \text{ (T in } {}^\circ K) \tag{12}$$

The free energy change is positive for all temperatures of interest

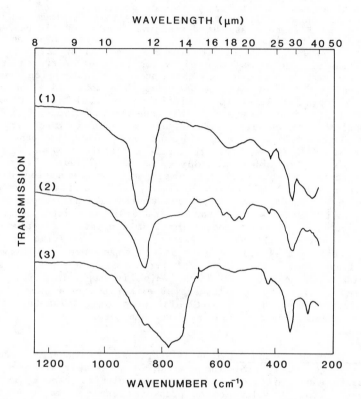

Figure 2. Infrared transmission spectra for: (1) Single crystal germanium with HPO formed GeO_2; (3) Specimen as in (1) after thermal nitridation in NH_3; and (2) Specimen as in (3) after HPO-Reoxidation.

and thus is not stable. Using this result and the equilibrium
constant for that reaction, it can be shown that to produce a
stable nitride from the elements, excessively high nitrogen
pressures would be required. Also, the existence of the nitride at
such temperatures and atmospheric pressure depends on a nitride
dissociation rate that is kinetically inhibited, since the free
energy change for the formation of the nitride is positive.

Wild et al. (122) introduced small amounts of oxygen into the
reaction chamber and thereby further stabilized the nitrides of
germanium at elevated temperatures. In varying the water
vapor/hydrogen ratio in an ammonia atmosphere, they were able to
prepare various mixtures of "α" and "β" germanium nitride by
controlling the P_{H_2O}/P_{H_2} ratio and the temperature. At 700°C, a
maximum amount of α-nitride was observed in the mixture, for a
constant oxygen partial pressure in the system. Based on this,
they postulated that α-nitride forms via a vapor phase mechanism,
most probably by the reaction of ammonia with volatile germanium
monoxide (which forms by the "active" oxidation of germanium at
these low oxygen pressures) according to the reaction.

$$11.5 \text{ GeO} + 15 \text{ NH}_3 = \text{Ge}_{11.5}\text{N}_{15}\text{O}_{0.5} + 11\text{H}_2\text{O} + 11.5\text{H}_2 \qquad (13)$$

When the equilibrium constant for this equation is combined with
the constants already established for Equations 8 and 10, the
resulting free energy of formation of the oxynitride is negative.
Therefore, the nitride will form under these conditions. At 700°C,
for example, Wild et al. estimated that the free energy change
associated with α-Ge$_3$N$_4$ formation was -109 kcal/mole, which is in
agreement with the values estimated by Kubashewski and Evans (129)
from heats and entropies of formation.

3.5. Formation Mechanisms of Thermally Grown Ge$_3$N$_4$. Although a
considerable amount of information on the formation mechanisms of
Si$_3$N$_4$ film growth has appeared in the literature (119,120,130,131),
very little exists on thermally grown germanium nitride. However,
based on the existing literature, some analogies from Si$_3$N$_4$
formation, and recent morphological evidence, a growth mechanism of
Ge$_3$N$_4$ formation can be proposed.

All thermodynamic and morphological evidence to date
(80,122,126) suggests that α-Ge$_3$N$_4$ forms from the vapor phase while
β-Ge$_3$N$_4$ forms via a solid state reaction; i.e. direct reaction
between solid Ge and dissociated nitrogen. When Wild et al. (122)
nitrided germanium wafers and powders in flowing NH$_3$, they found
the α-phase always formed at the gas-solid interface and the β-
phase formed at or near the nitride-semiconductor interface.
During the initial stages of nitridation, when "bare" Ge surfaces
are exposed, the nitride layer consists almost entirely of α-Ge$_3$N$_4$.
(This is consistent with similar morphologic observations made on
silicon wafers that were nitrided in flowing N$_2$:H$_2$ mixtures
(119,120,130).) They (122) attribute the vapor phase formation of
α-Ge$_3$N$_4$ to the appreciable vapor pressure of GeO established over
the germanium. Since the vapor pressure of germanium is very small
relative to that of germanium monoxide (10^{-14} atm vs. 10^{-5} atm at
700°C), the formation of α-Ge$_3$N$_4$ by reaction with GeO is more
likely than reaction with elemental germanium. Popper and

Ruddlesden (111) determined that heating germanium powder in pure ammonia at 750°C yielded a product that was almost all α-phase whereas heating germanium dioxide under the same conditions produced a predominant β-layer. Gilles (132) and Labbe (133) confirmed the results of Popper and Ruddlesden and they postulated that the β-phase was stabilized by an oxide, oxygen, or steam, since the proportion of the β-phase in the product mixture was increased when only pure ammonia was used to react with the germanium. In yet another reexamination of these results, Remy and Pauleau (126) investigated the formation of Ge_3N_4 by three different methods including CVD and concluded that oxygen is not necessary for the formation of either crystalline phase. The α-phase could be formed directly by the reaction of ammonia and germanium in low ambient oxygen partial pressures supporting the observations made by Wild et al. (122). They (126) also observed that during the initial stages of nitridation, "bare" Ge surfaces exposed to ammonia quickly become covered with an α-Ge_3N_4 layer.

The summation of the various observations described in the preceeding paragraph leads to the conclusion that the nature of the residual surface "native" oxide strongly affects the subsequent phase that forms. If GeO is present then an active molecular species is available in the near-surface gases to promote α-Ge_3N_4 formation and subsequent condensation. On the other hand, if GeO_2 is present a passive surface condition exists requiring nitridation to proceed via the solid diffusion of a nitridant into the surface. Furthermore if a uniform passive oxide layer is present on the surface, subsequent α-nitride can only be produced in the vicinity of oxide disruptions.

The nitridant also plays an important role in the nitridation of germanium and germanium oxide to form germanium nitride. It is generally agreed that Ge_3N_4 cannot be formed directly from the elements, primarily because molecular nitrogen is so unreactive. The thermal dissociation of molecular nitrogen to form atomic nitrogen is so thermodynamically unfavorable that the equilibrium concentration of atomic nitrogen at one atmosphere total pressure is exceedingly small, even at elevated temperatures (134). However, similar calculations and measurements for the dissociation of ammonia (Equation 11) show a substantial concentration of atomic nitrogen at elevated temperatures (126,134). For example, from (126), only 0.4% ammonia was dissociated at 630°C as compared to more than 20% dissociation at 800°C. All this suggests that NH_3 is a more effective nitridant than N_2, particularly under conditions where considerable diffusion has to occur prior to reaction, as is the case when a native oxide exists on the surface. Remy and Pauleau (126) established that higher temperatures favored β-phase growth regardless of the deposition method used, which related to the decomposition of ammonia, the production of greater amounts of active nitrogen, and the enhanced diffusion rates.

Adding hydrogen to the ammonia alters the equilibria established by Equation 3 and thus can be used to control the amount of atomic nitrogen formed by dissociation. Remy and Pauleau (126) were able to prepare essentially pure α- and β-Ge_3N_4 using a mixture of 65% NH_3 and 35% H_2 at 630°C and 800°C, respectively.

It has been demonstrated by FIR-transmission analysis (72), that germanium oxide readily reacts with ammonia to form nitride phases but is merely reduced by $N_2:H_2$ mixtures. Even though in both instances hydrogen was available to dissolve the Ge-O bonds, the atomic nitrogen produced from NH_3 dissociation is evidently also necessary for the conversion to occur. In thick oxides, atomic nitrogen might also react with free or incompletely bound germanium in the oxide or at the oxide-germanium interface. This was experimentally verified by FIR transmission measurements (72,80). In those experiments, the well established α- and β-nitride absorption bands were observed to appear immediately and thereafter intensify with nitridation time. The residual oxide peak was observed through the course of the experiments. This was quite different from the results described above for Stein's ion-implanted FIR measurements (103) where a single transmission minimum was observed to shift continuously from the oxide position to the nitride position as a function of O/N ratio in the implant.

In a recent review by Jennings (135) on reactions between silicon and nitrogen, the difference between the Si_3N_4 phases was explained in terms of competing reactions. The reaction to form α-Si_3N_4 was attributed to the complexing of molecular nitrogen with silicon whereas the reaction to form β-Si_3N_4 was attributed to the complexing of atomic nitrogen with silicon. Thus far, all evidence suggests that this is also the case for Ge_3N_4 formation.

In gas-solid reactions, such as described here. The transport of reactants to the solid surface or removal of products from the surface is frequently the rate controling step. The thermal nitridation of GeO in flowing ammonia to form α-Ge_3N_4 can be modeled in this manner. Wild et al. (122) introduced the concept of a hydrodynamic boundary layer to explain the formation of α-Ge_3N_4 from the vapor phase. In their model, NH_3 and H_2O vapor were transported to the Ge surface and volatile GeO was transported from the Ge surface through a single boundary layer. Surface decomposition was achieved via the oxidation of Ge activated by H_2O vapor. The gradient of the various species through the boundary layer provides the driving force for mass transfer through the layer. For each equilibrium that is established, the thickness of the boundary layer can be adjusted by changing the gas flow (136). If boundary layer thickness is reduced by increasing gas flow, the diffusion length of the nitriding species in the boundary layer is reduced and the rate of α-Ge_3N_4 formation is enhanced. However, if the gas flow rate is increased beyond a critical value the volatile GeO will be carried away from the vicinity of the germanium surface and not participate in the reaction. A quantitative evaluation of this critical flow rate can be estimated from the Peclet number (Pe): the ratio of the convective flux to the diffusive flux (137). Under conditions where Pe > 1, the convective flux dominates and relatively little α-Ge_3N_4 will form on the surface and β-Ge_3N_4 growth will be favored. Therefore, the relative amounts of α- and β-nitride formed on germanium can be adjusted by adjusting the ammonia flow rate.

Microstructural observations have provided further evidence that the α-phase forms from the condensation of a vapor. Compos-Loriz and Riley (138) and Jennings (139) have proposed that α-Si_3N_4 forms from reactions involving SiO, probably by a vapor-liquid-solid mechanism which tends to form needles or protruding crystals with high aspect ratios. It has also been observed on nitrided

silicon crystals that these same needles form adjacent to breaks
and fissures in the nitride layer where SiO can volatilize
(120,139). Also, α-Si₃N₄ invariably forms when the nitride is
prepared by chemical vapor deposition (123,124). On the other
hand, β-Si₃N₄ growth results in a morphologically dense, coherent
layer comprised of well formed hexagonal crystallites (135,139).
Beta-phase formation is therefore attributed to the direct reaction
between nitrogen and solid silicon.

In several recent studies involving the nitridation of
germanium and germanium dioxide (72,80,101,126), many of the same
formation mechanisms were observed. Hua et al. (101) observed the
deposition of a nitride product in what appeared to be etch pits
and polishing defects. They suggested that these defects are
produced from contamination of oxygen in the reactor and act as
nucleation sites for crystallization of the nitride. More recent
morphological studies of the nitridation of germanium oxide (80)
have shown that these are thermal etch pits which arise from
elevated processing temperatures and act as sites for the
preferential volatilization of germanium or GeO which subsequently
reacts with ammonia and redeposits in the vicinity of the defect.
Experiments with different gas flow rates have shown that the
growth of nitride in these defects can be varied. After etching
the germanium surface in a NaOCl:HF:H₂O solution, Hua et al. (101)
found that the preferential crystallization of the nitride did not
occur and smooth uniform layers resulted. The etching most likely
reduced any native oxide, thereby limiting the sublimation of GeO
and its subsequent α-forming reaction. A similar effect was
observed when germanium dioxide layers were nitrided in ammonia
under high flow rate conditions (80). A relatively dense, uniform
nitride layer resulted under these conditions which was
characterized as β-Ge₃N₄. The difference in nitride surface
morphologies produced under different conditions is shown in Figure
3b and 3c along with the initial oxide morphology, 3a. Rosenberg
(7,101) has observed that uniform films grow rapidly on the surface
of germanium in an ammonia ambient but are frequently accompanied
by anisotropic vapor etching which is responsible for the observed
surface defects. Dalgleish et al. (119,130) have also reported a
similar nitride growth on single crystal silicon which emanates
from the thermal etch pits, characteristic of preferential
volatilization. Therefore, the density of thermally generated
defects and mechanically induced defects have a strong effect on
the nature of the nitride produced.

3.6. Kinetics of Ge₃N₄ Growth. As with the growth of Si₃N₄ on
silicon, the growth of Ge₃N₄ on germanium by thermal nitridation is
limited by the formation of a product layer which quickly covers
the surface of the semiconductor and separates the reacting
species. From this behavior, the anticipated growth kinetics
should be parabolic, after the initial coverage occurs, since
further reaction is limited by diffusion of rectants through the
nitride. However, the kinetics governing this process are more
complex due to the competing growth of two different nitride phases
which form by different mechanisms and under different reaction
conditions. A further complication is introduced by the
possibility of amorphous nitride formation.

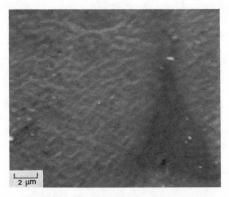

Figure 3a. Electron micrograph of germanium oxide formed by HPO on single crystal germanium (111) surface.

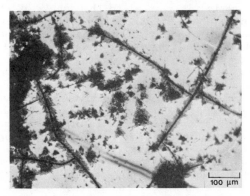

Figure 3b. Optical micrograph of HPO GeO_2 film on (111) germanium surface nitrided under low ammonia flow condition.

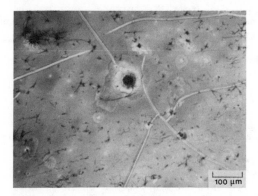

Figure 3c. Optical micrograph of HPO GeO_2 film on (111) germanium surface nitrided under high ammonia flow condition.

Since the two different crystalline nitride phases form by different mechanisms the respective reaction rates should be governed by different rate laws. Little data is available on the growth rates of the individual nitride phases on single crystal germanium. The only kinetic data available are those recently reported by Rosenberg (7) and Hua et al. (101), which includes the amorphous formation as well. However, the growth kinetics of Si_3N_4 on silicon have been extensively studied (140,141,142,143,144) and some inferences can be drawn from this system. Longland and Moulson (143) have isolated the individual α- and β-nitride reaction kinetics during the initial nitridation of a Si surface and found them to be identical. When nitridation is continuous, the formation reaction slows down as a dense nitride forms on the surface and the α-formation reaction continues. This led them to conclude that the nitridation of a Si surface leads to α-Si_3N_4 formation. Both growth kinetics, however, are nearly parabolic, suggesting that a protective layer eventually forms in either case, limiting diffusion to the interface.

The rate controlling mechanism associated with the initial linear growth kinetics is related to the supply of atomic nitrogen to the reaction site (131). As the nitride nuclei form on the silicon surface, they deplete the adjacent regions of the Si surface of chemisorbed nitrogen. The nuclei grow laterally as a result of surface diffusion and eventually approach one another, coalescing into a uniform surface layer. At this point the kinetics change due to the limited supply of silicon to the outer surface and nitrogen to the inner surface of the nitride layer. Other investigations (145,146) into the nitridation kinetics of silicon wafers have shown that there is an initial rapid growth of nitride followed by inhibited growth which gives a logarithmic rate for crystalline nitrides. This process can occur several times during the course of the reaction and has been related to the build up of a thick film, which eventually cracks due to crystallization events, allowing the nitrogen free access to the silicon surface and the process is repeated. In summation then, the growth kinetics can be directly related to the diffusional processes. When the nitride film thickness is less than the diffusion length of the nitridant, the growth is surface reaction limited and linear kinetics are dominant. However, as the film thickness increases and becomes greater than the diffusion length of the nitridant, growth is diffusion limited and a logarithmic rate law results, having the form:

$$thickness = K \ln (1 + a \cdot time) \qquad (14)$$

where K and a are constants.

Hua et al. (101) observed similar behavior during the nitridation of germanium wafers in ammonia ambients with some O_2 present. Their data shows rapid initial germanium nitride growth followed by a period of slow growth. They also observed an increase in growth rate for long times which is most likely related to the rupturing of the nitride film followed by increased surface reaction, as already mentioned. Since they also observed time varying nitrogen contents in the films, they proposed that the growth mechanism involves the diffusion of nitrogen into and through the film which eventually becomes the rate limiting step for additional growth. Hua et al. attributed the increase in rate at long times to the crystallization of the initially amorphous, nitride layers. It has been well documented by Hedvall (147) that solid state reaction may be accelerated at a polymorphic transition point. At such transitions, large volume changes can occur which are responsible for lattice strains and enhanced material transport. Therefore, any structural changes or strains in the nitride as it forms can affect the growth rate of the nitride by changing the diffusion of nitrogen species through the film.

3.7. Preparation and Properties of Germanium Nitride Films.
Thermally grown germanium nitride films have been discussed in great detail due to the ease with which such films could be incorporated into the conventional processing of integrated circuits. In this section that follows, some of the properties of thermal nitrides of germanium including index of refraction, specific resistance, dielectric constant and breakdown voltage are reported. A summary of the preparation and properties of germanium nitride films deposited by other methods is included.

The refractive index (η) of a film, determined by ellipsometry, is frequently used as a measure of the extent of nitridation in the film. Thermal nitrides of germanium prepared by the nitridation of germanium in oxygen bearing ammonia ambients (7,101), resulted in refractive index values from 1.5 to 2.3 depending on growth conditions and NH_3/H_2 ratio. When pure NH_3 was used, refractive indices greater than 2.0 were realized. During the nitridation of thick germanium oxide films (72), we observed a change in the index of refraction from 1.7 to 2.1 depending on the degree of conversion of the oxide to the nitride. After long times at elevated temperatures (700°C) the value for the index of refraction remained at 2.1. Refractive indices of thin amorphous films produced by chemical vapor deposition (107,149) are between 2.00 and 2.06 depending on deposition condition. Bagratishvili et al. (109) obtained similar values for the refractive index of nitride films produced by the direct reaction between hydrazine and germanium. These values, along with the refractive index of 1.61-1.69 established for GeO_2, set the limits within which the refractive indices of all oxynitride and nitride films fall. As the nitrogen content in the nitride increases, the refractive index approaches the upper limit. When the nitride films produced by the nitridation of thick oxides were reoxidized (72), we observed that the index of refraction of the film approaches 1.65 as the re-oxidation of the film occurs.

Germanium oxynitride films have been prepared by the reactive sputtering of germanium in a hydrazine plasma (104,105,106) and compositional analysis of the resulting films showed that they were nonstoichiometric. Appreciable amounts of oxygen were incorporated into those films due to the oxygen partial pressure established in the plasma (1.3×10^{-2}-6.7×10^{-3} Pa). The oxygen contents in these films of the form $Ge_3N_{4-x}O_{1.5x}$ ranged from 2.0-14.0% and the corresponding refractive indices varied linearly from 3.43 to 2.20. This continuous variation of index of refraction shows that substoichiometric germanium nitride films were formed by reactive sputtering: i.e. excess germanium must exist in these films.

The nitride thin films formed on germanium under various conditions exhibited substantially slower etch rates in various acid solutions and water than the thermal oxides formed on germanium. Nagai and Niimi (107) found that nitride films produced by CVD were only slightly soluble in concentrated HCl, H_2SO_4, H_3PO_4 and NaOH solutions whereas rapid dissolution occurred when placed in concentrated HF and HNO_3 solutions. However, they did establish that the etch rate in HF or HNO_3 solutions decreased with increasing growth temperatures, similar to the etching behavior of CVD-Si_3N_4 films. Bagratishvili et al. (109) obtained similar results on nitride films prepared by reaction with hydrazine. When germanium nitrides were prepared in our laboratory (72,80) by the nitridation of thick GeO_2 films the films were virtually insoluble in deionized water, indicating that those oxides were stabilized by nitriding treatments.

The value of the dielectric constant of nitride films were also considerably greater than the values obtained for thermal oxides. Yashiro (149) determined that the dielectric constant of vapor deposited Ge_3N_4 was independent of growth temperature and had average value of approximately 8.0. When nitride films were deposited by reactive sputtering (104,106), the dielectric constant was found to vary linearly with oxygen content in the films. Dielectric constants between 9 and 12 were obtained and those films with the least incorporated oxygen had the largest dielectric constant. Such behavior suggests that nitrides are intrinsicly better insulators than oxides and in that respect are more suitable for MOS devices and antireflection coatings.

The breakdown of an MOS device is a localized phenomenon caused by current concentration at low energy barrier points in the film. When the current density at such points increases beyond an allowable value, the device is destroyed. Therefore, film uniformity is critical to the success of such devices. The chemical properties of the nitrides formed on germanium in conjunction with the resulting surface morphology dictate its usefulness; i.e. the extent of crystallinity and distribution of phases in the nitride. To date, it is difficult at best to produce uniform nitrides, via thermal reaction, over large areas of the germanium crystal surface. Hua et al. (101), Rosenberg (7) and Gregory et al. (80) have identified thermal defects, mechanical defects, breaks, crystallization (or growth) defects and other discontinuities in the nitride films which lead to extremely low breakdown voltages. Hua et al. (101) reported a dielectric strength of $3x10^6$ V/cm for smooth amorphous nitride layer formed by thermal reaction. Crisman (35) measured the dielectric strength of nitrided germanium oxide films and reported an average value of 1.4×10^6 V/cm for films that were approximately 2200 A thick.

Slightly lower breakdown voltages were reported for vapor deposited Ge_3N_4 films ($8.5x10^5$ V/cm). Yashiro (149) also noted that in biased temperature measurements using C-V plots, the displacement of the C-V curve was opposite to that of ion-migration indicating that mobile Na^+ ions do not migrate readily in nitride films. Trapping of carriers in the nitride insulator was inferred as the mechanism of C-V curve displacement. Bagratishvili et al. (148) observed different behavior for nitride films produced by

reactive sputtering.   No hysteresis in the C-V curves was observed
suggesting that few traps were present in the insulator.   They also
noted that the breakdown voltage varied in the resulting MIS
structures from 2-7x10^6 V/cm depending on stoichiometry of the
predominantly amorphous films.   Crisman (35) has shown that even
poor quality oxide-semiconductor interfaces (as defined by C-V
curves) can be improved to an electronically useful range by
thermal nitridation.   The effect of such a treatment on the
properties of the insulator-semiconductor interface is shown in
Figure 4.   There is minimal hysteresis and curve "skewing" in the
C-V trace for the nitrided oxide whereas the oxide C-V trace shows
considerable deformation.   This implies that there are many more
"slow" interface states and possibly trapped charges as well in the
oxide prior to nitridation.
        Yashiro (149) also studied the properties of the Ge$_3$N$_4$/Ge
interface and found that the magnitude of the hysteresis and the
interface state density (of C-V curves corresponding to vapor
deposited Ge$_3$N$_4$ films) decreased with increasing deposition
temperature up to 550°C.   This was attributed to the number of
unsatisfied germanium bonds at the interface when the Ge$_3$N$_4$ films
were deposited at lower temperatures.   Apparently, during
deposition, the active nitrogen derived from the ammonia reacted
with the unbound germanium satisfying these bonds and reducing the
hysteresis. Above 550°C, an opposite trend was observed, suggesting
that unsatisfied bonds were being established as a result of the
thermal processing.   The effect of the annealing on the interface
properties of these vapor deposited films was also studied.   It was
found that the hysteresis decreased with increasing annealing
temperature while the interface state density increased with
increasing temperature.   This apparent anomalous behavior is not
well understood at this time.
        A summary of the electrical properties of Ge$_3$N$_4$ films prepared
by a variety of methods is presented in Table IV.   The pertinent
chemical reactions corresponding to the different preparation
techniques is shown in Table V, and includes approximate processing
temperatures for the different methods.

Summary

        Because of the usefulness of silicon oxides and silicon
nitrides as processing tools in the semiconductor industries, a
large volume of information has been generated on these materials.
The studies of germanum insulators have been limited in scope,
dealing primarily with the thermodynamics and kinetics of the
reactions between germanium and oxygen/nitrogen.   Nevertheless,
significant advances have been reported for those materials with
respect to the electronic quality of the insulator films and their
interfaces with single crystal germanium surfaces.   Based on our
review of the recent literature, we conclude the following points
are significant to future development of germanium based device
technology:

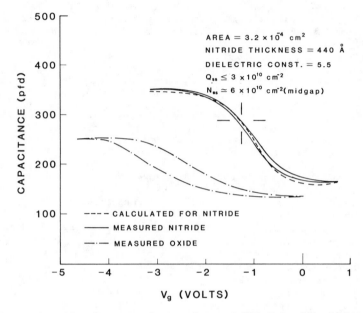

Figure 4. Capacitance-voltage plots of HPO-GeO$_2$ film before and after nitridation in high flow ammonia. Calculated C-V curve is provided for comparison.

Table IV. Properties of Ge_3N_4 Films Prepared by Different Techniques

Property	Direct Thermal	CVD Ge_3N_4 ([107],	Thermal in Hydrazine	Reactive Sputtering ([104],[106])	Thermal nitrid- ation of GeO_2 ([35])
Structure	amorphous/ crystalline	amorphous	amorphous	amorphous	crystalline
Index of refraction	1.5-2.3	2.0-2.06	2.06	2.2-3.43	2.7-2.8
Dielectric Constant	-------	8.0	6-8	9.0-12.0	5.5-6.9
Dielectric Strength (V/cm)	$3x10^6$	$8.5x10^5$	$5.6x10^6$	$2-7x10^6$	$1.4x10^6$
Specific Resistance (ohm-cm)	-------	$10^{14}-10^{16}$	$5x10^{13}$ $5x10^{14}$	------	------
Fixed Surface Charge Density (# cm^{-2}	-------	$1x10^{15}$	------	$3-5x10^{10}$	$3x10^{10}$
Interface State Density (# $cm^{-2}eV^{-1}$)	------	$1-4x10^{12}$	------	$1x10^{11}$	$5x10^{10}$

Table V. Preparation Techniques for Ge_3N_4 Films

Preparation Method	Chemical Reaction	Temperature °C	Reference
Nitridation of Germanium wafers	$3Ge+4NH_3=Ge_3N_4+6H_2$	600–700	(7,101, 102)
Nitridation of Germanium Wafers	$3Ge+2N_2H_4=Ge_3N_4+6H_2$	650	(109)
Nitridation of GeO_2 Films	$3GeO_2+4NH_3=Ge_3N_4+6H_2O$	600–750	(72)
Nitridation of Native Ge Films	$3Ge+4NH_3=Ge_3N_4+6H_2$	350–450	(101,102)
Chemical Transport	$3Ge+2N_2H_4=Ge_3N_4+6H_2$	350–450	(109)
Chemical Vapor Deposition	$3GeCl_4+4NH_3=Ge_3N_4+6H_2$	200–400	(107,108 109)
Reactive R.F. Sputtering	$3Ge+2N_2H_4=Ge_3N_4+6H_2$	20–400	(104,106)
Ion Implantation	$Ge+N_2+(N^+) = Ge_3N_4$	200	(103)
Microwave Excitation	$3N+3GeH_4=Ge_3N_2+6H_2+\frac{1}{2}N_2$	100–200	(110)
Microwave Excitation	$3N+3GeCl_4=Ge_3N_2+6Cl_2+\frac{1}{2}N_2$	65	(110)
Electron Beam Evaporation	$Ge_3N_4 = Ge_3N_4$	350	(148)

(1)   As with silicon and oxygen, germanium and oxygen form an amorphous monoxide, an amorphous dioxide and several crystalline dioxide phases with major and minor structural modifications.

(2)   Unlike silicon, the monoxide phase is volatile at the temperature of its formation from elemental components and, therefore tends to make direct thermal oxidation difficult as a passivation and masking technique.

(3)   Neither the stability nor the identification of the monoxide phase in thin films has been well established. (Most oxides to date have been prepared as mixtures of oxide phases.)

(4)   The monoxide is, however, metastable at room temperature, and therefore, chemical surface preparations can result in active surfaces: particularly, important when subsequent nitridation is anticipated.

(5)   The cleaning and etching procedures strongly affect subsequent oxide and nitride growth morphology and kinetics. Therefore a better understanding of these procedures needs to be developed.

(6)   Regardless of pre-oxidation on pre-nitridation surface preparation, the starting surfaces of any germanium crystal will contain 10-100 A of a native oxide. The exact nature of surface oxides, i.e. activity, phase, and thickness, will depend on the preparation history.

(7)   Thick surface films of either amorphous-GeO_2 or hexagonal-GeO_2 can be prepared by HPO.

(8)   The formation of tetragonal-GeO_2 appears to be desirable for chemical stability and reduction of interface strain deformation. The tetragonal phase can grow epitaxially on the (110) plane of germanium and therefore is a promising candidate for MOS structures, but conditions for its formation are not well defined.

(9)   Interfaces with low density of states can be formed by oxidation of germanium through SiO_2 masks or by direct, high pressure, thermal oxidation. Silica masked interfaces are superior at present, possibly because the variation in parameters for HPO has not yet been extensively studied.

(10)   Thick germania films of a predominantly amorphous nature appear to be a reasonable starting point for the formation of germanium nitrides.

(11) Alpha-Ge_3N_4 is formed on Ge surfaces by a vapor phase reaction and most likely by the nitridation of volatile germanium monoxide whereas β-Ge_3N_4 is formed by reaction with incompletely bound germanium in the oxide or at oxide-semiconductor interfaces and requires substantial solid state diffusion of nitridant to effect the conversion.

(12) Beta Ge_3N_4 growth results in uniform, dense and defect free passivation layers whereas α-Ge_3N_4 results in non-uniform-defected films.

(13) Lowering the reaction temperature reduces defect generation and thus produces more uniform nitride layers.

(14) Both composition and flow rate of the nitriding ambients strongly affect the morphology of the nitride films with the presence of nacent nitrogen and possibly hydrogen as well being particularly important for β-nitride formation.

(15) Heat treatment of HPO-GeO_2 in ammonia ambient results in low interface state density even for poor quality oxides as the starting material.

(16) Processing techniques for thermally grown oxides and nitrides can be incorporated into existing planar device technology.

Acknowledgments

The authors would like to thank Mary Costa, Kevin Fish and Catherine Gavin for the timely preparation of this manuscript. In addition we also acknowledge the Materials Research Laboratory at Brown University and the Chemical Engineering Department at the University of Rhode Island for support of this work.

Literature Cited

1. T. Torikai, I. Hino, H. Iwasaki and K. Nishida, Jap. J. of Appl. Phys. 21, p. L776 (1982).
2. Infrared Detectors, edited by R. D. Hudson and J. W. Hudson, Dowden, Hutchinson & Ross, Inc. Stroudsburg, Pennsylvania (1975).
3. V. F. Synorov, A. Antoniskis, A. I. Chernyshev and N. M. Aleinikov (Electrical Properties of a Passivated Germanium Surface) Elektron. Protsessy. Poluprov. Tr. Simp., p. 153 (1967).
   - A. A. Frantsuzov and N. I. Makrushin (Absorption of Hydrogen and Oxygen on the Surface of Single Crystal Germanium), Electron. Protsessy. Poverkhn. Poluprovodn. Granitse Razdela Poluprovodn. - Dielektr. Mater. Simp. 5th, p. 257 (1974).
   - Y. I. Belyakov and T. N. Kompaniets, (Adsorption of Oxygen and Germanium Dioxide on a Germanium Surface) Zh. Terk. Fiz. 43 (6), p. 1317 (1973).

4. J. Kiraly, (Surface Oxidation Phenomena on a Germanium
   Single-Crystal Surface. IV. Oxidation Mechanisms) Masy. Kem.
   Foly. 76 (5), p. 217 (1970).
   - J. Kiraly and P. Fejes (Surface Oxidation Phenomena on a
   Germanium Single-Crystal Surface. III. Effect of
   Temperature on Sorption Rates) ibid. 74 (12), p. 571, (1968).
   - J. Kiraly (Surface Oxidation Phenomena on a Germanium
   Single-Crystal Surface. II. Oxygen Absorption Kinetics)
   ibid. 74 (11), p. 553 (1968).
5. T. Yashiro, H. Nagai and K. Yano (Passivation of Germanium
   Devices. II. Electrical Properties of Germanium-Silicon
   Dioxide (SiO_2) and German-Silicon Nitride (Si_3N_4) Interfaces)
   Denki Tsushin Kenkyujo Kenkyu Jitsuyoka Hokuku 20 (4), p. 927
   (1971).
   - T. Yashiro and T. Niimi (Formation of Germanium Dioxide
   Films for Passivation of Germanium Devices) ibid. 17 (12), p.
   2987 (1968).
   - T. Shibata, Y. Igarashi and K. Yano (Passivation of
   Germanium Devices. III. Fabrication and Performance of
   Germanium Planar Photodiodes) ibid. 20 (4), p. 965 (1971).
6. E. H. Nicollian and J. R. Brews, MOS Physics and Technology,
   John Wiley and Sons, New York, NY (1982).
7. J. Rosenberg, Proceedings of Device Research Conference,
   Burlington, Vermont (June, 1983) to be published in IEEE
   Trans. Elect. Devices.
8. M. Green, Progress in Semiconductors, Volume 4, p. 35, edited
   by A. F. Gibson, John Wiley & Sons, Inc., New York, NY
   (1960).
9. A. J. Rosenberg, J. N. Butler and A. A. Menna, Surface
   Sciences 5, p. 1 (1966).
10. M. Green, J. A. Kafalas and P. H. Robinson, Semiconductor
    Surface Physics, p. 349, edited by R. H. Kingston, University
    of Pennsylvania Press, Philadelphia, PA (1957).
11. J. A. Dillon and H. E. Farnsworth, J. Appl. Phys. 28, p. 174
    (1957).
12. R. E. Schlier and H. E. Farnsworth, Semiconductor Surface
    Physics, p. 3, edited by R. H. Kingston, University of
    Pennsylvania Press, Philadelphia, PA (1957).
13. R. Ludeke and A. Koma, Phys. Rev. Letters 34, p. 1170 (1975).
14. R. F. Lever, Surface Science 9, p. 370 (1968).
15. J. von Wienskowski and H. Froitzheim, Phys. Stat. Sol. (b)
    94, p. 429 (1979).
16. Private communication, R. Schwartz, Electrical Engineering
    Department, Purdue University.
17. C. R. Helms, W. E. Spicer and V. Pereskokov, Appl. Phys.
    Letters 24, p. 318 (1974).
18. V. E. Henrich and J. C. C. Fan, J. Appl. Phys. 46, p. 1206
    (1975).
19. S. G. Ellis, J. of Appl. Phys. 28, p. 1202 (1957).
20. M. Green and J. A. Kafalas, Electrochemical Society Meeting,
    Cincinnati, Ohio (May 1955) and Phys. Rev. 98, p. 1566
    (1955).
21. A. A. Frantsuzov and N. I. Makrushin, Surface Sciences 40, p.
    320 (1973).

22. R. J. Madix, R. Parks, A. A. Susu and J. A. Schwarz, Surface Science 24, p. 288 (1971).
23. J. R. Story, J. Electrochem. Soc. 112, p. 1107 (1965).
24. Y. Margoninski and L. G. Fienstein, Physics Letters 31A, p. 153 (1970).
25. D. T. Stevenson and R. J. Keyes, Physica 20, p. 1041 (1966).
26. R. H. Kingston, J. Appl. Physics 27, p. 101 (1956).
27. Y. M. Ercil, "A Study of Passivation of Germanium by High Pressure Oxidation," Ph.D. Thesis, Brown University, Providence, RI, (1981).
28. F. R. Archer, J. Electrochem. Soc. 104, p. 619 (1957).
29. R. E. Schlier and H. E. Farnsworth, J. Appl. Phys. 34, p. 1403 (1959).
30. H. E. Farnsworth, R. E. Schlier, T. H. George and R. Berger, J. Appl. Phys. 29, p. 1150 (1958).
31. G. A. Barnes and P. C. Banbury, Proc. Phys. Soc. 71, p. 1020 (1958).
32. F. Allen and A. B. Fowler, J. Phys. Chem. Solids 3, p. 107 (1957).
33. I. V. Rzhanov and I. G. Neizvestny, Thin Solid Films 58, p. 37 (1979).
34. M. D. Jack, J. Y. M. Lee and H. Lefevre, J. of Electronic Materials 10, p. 571 (1981).
35. E. E. Crisman, "The Construction and Characterization of Native Insulators on Gallium Arsenide and Germanium", Ph.D. Thesis, Brown University, Providence, RI (1984).
36. A. Many and D. Gerlich, Phys. Rev. 107, p. 404 (1957).
37. W. H. Brittain and J. Bardeen, Bell System Technical Journal 32, p. 1 (1953).
38. W. Eitel, Structural Conversions Systems and Their Importance for Geological Problems, paper 66, Geolog. Soc. of Am., New York, NY (1958).
39. H. Statz, G. A. de Mars, L. Davis and A. Adams, Phys. Rev. 101, p. 1272 (1956).
40. J. Giber, L. E. Czran and M. Wegner, Acta. Khimica. Academiae Scientiarum Hungaricae 68, p. 279 (1970).
41. C. Winkler, J. Prakt. Chem. 142, p. 177 (1886).
42. L. M. Dennis and R. E. Hulse, J. Am. Chem. Soc. 52, p. 3553 (1930).
43. O. H. Johnson, Chem. Review 51, p. 431 (1952).
44. R. B. Bernstein and D. Cubicciotti, J. Am. Chem. Soc. 73, p. 4112 (1951).
45. Constitution of Binary Alloys. 1st Supplement, edited by Elliot, p. 483, McGraw-Hill, New York, NY (1969).
46. Michael Hoch and H. L. Johnston, J. Chem. Phys. 22, p. 1376 (1954).
47. W. J. Jolly and W. M. Latimer, J. Am. Chem. Soc. 74, p. 5757 (1952).
48. J. T. Law and P. S. Meigs, J. Electrochem. Soc. 104, p. 154 (1957).
49. J. T. Law and P. S. Meigs, Semiconductor Surface Physics, p. 383, University of Pennsylvania Press, Philadelphia, PA (1957).
50. H. Bohm, Naturwissenschaften 55, p. 648 (1968).

51. D. Yamaguchi, K. Kotera, M. Asano and K. Shimizu, J. Chem. Soc. Dalton Trans., p. 1907 (1982).
52. ASTM card nos. (H) 4-0497, (T) Rutile 9-379, (T) Cristobalite 21-902, (C) 23-999.
53. K. J. Seifert, H. Nowotny and E. Hauser, Monatsh. Chem. 102, p. 1006 (1971).
54. E. R. Lippencott, A. Van Valkenburg, C. E. Weir and E. N. Bunting, J. of Research of N.B.S. 61, p. 2885 (1958).
55. M. Stapelbroek and B. D. Evans, Solid State Communications 25, p. 959 (1978).
56. F. J. Arlinghaus and W. A. Albers, J. Phys. Chem. Solids 32, p. 1455 (1971).
57. J. Sladkova, Czech. J. Phys. B18, p. 801 (1968).
58. A. A. Frantsuzov and N. I. Makrushin, Soviet Physics - Semiconductors 5, p. 619 (1971).
59. A. W. Laubengayer and D. S. Morton, J. Am. Chem. Soc. 54, p. 2303 (1932).
60. J. Sladkova, Czech. J. Phys. B27, p. 943 (1977).
61. E. E. Crisman, Y. M. Ercil, J. J. Loferski and P. J. Stiles, J. Electrochem. Soc. 129, p. 1845 (1982).
62. W. D. Kingery, Introduction to Ceramics, John Wiley and Sons, New York, NY (1960).
63. J. H. Muller and H. R. Blank, J. Am. Chem. Soc. 46, p. 2358 (1924).
64. J. F. Sarvar and F. A. Hummel, J. Am. Chem. Soc. 43, p. 336 (1960).
65. E. S. Candidus and D. Tuomi, J. Chem. Phys. 23., p. 588 (1955).
66. F. A. Trumbore, C. D. Thurmond and M. Kowalchik, J. Chem. Phys. 24, p. 1112 (1956).
67. L. Baewea and P. Zavitsanos, Phys. Chem. Solids 2, p. 284 (1957).
68. Y. Kotera, M. Yonemura, Trans. Faraday Soc. 59, p. 147 (1963).
69. W. A. Albers, Jr., E. W. Valyocsik and P. V. Mohan, J. Electrochem. Soc. 113, p. 196 (1966).
70. T. Yashiro, Y. Igarashi and T. Niimi, Jap. J. Appl. Phys. 7, p. 556 (1968).
71. M. K. Murthy and E. M. Kirby, Phys. Chem. of Glasses 5, p. 144 (1964).
72. E. E. Crisman, O. J. Gregory and P. J. Stiles, J. Electrochem. Soc. 131, p. 1896 (1984).
73. G. V. Samsonov, Oxide Handbook, p. 333, IFI/Plenum Press, New York, NY (1973).
74. R. J. Zeto, C. G. Thornton, E. Hryckowian and C. D. Bosco, J. Electrochem. Soc. 122, p. 1409 (1975).
75. B. E. Deal and A. S. Grove, J. Appl. Phys. 36, p. 3770 (1965).
76. Private communication from J. Severns, Naval Research Laboratory.
77. E. S. Vera, J. J. Loferski, M. Spitzer and J. Schewchin, Third European Communities Photovoltaic Solar Energy Conference, p. 911, Cannes, France (1980).
78. M. Lasser, C. Wysocki and B. Bernstein, Phys. Rev. 105, p. 491 (1957).

79. M. Zerbst, Z. Angew Phys. 22, p. 30 (1966).
80. O. J. Gregory, E. E. Crisman and P. J. Stiles, AICHE Symposium on Interfacial Phenomenon in the Semiconductor Industry, San Francisco, California (November, 1984).
81. M. Kuisl, Solid State Electronics 15, p. 595 (1972).
82. I. Franz and W. Langheinrich, Solid State Electronics 13, p. 807 (1970).
83. K. L. Wang and A. Joshi, J. Vac. Sci. Technol. 12, p. 927 (1975).
84. K. L. Wang and P. V. Gray, J. Electrochem. Soc. 123, p. 1392 (1976).
85. K. K. Schroder, Appl. Phys. Letters 25, p. 747 (1974).
86. H. J. Stein, J. Electrochem. Soc. 121, p. 1073 (1974).
87. B. R. Appleton, O. W. Holland, J. Narayan, O. E. Schow, III, J. S. Williams, K. T. Short and E. Lawson, Appl. Phys. Letters 41, p. 711 (1982).
88. O. W. Holland, B. R. Appleton and J. Narayan, Oak Ridge National Laboratory, (unpublished).
89. F. L. Edelman, L. N. Alexandrov, L. I. Fedina and V. S. Latuta, Thin Solid Films 34, p. 107 (1976).
90. E. W. Valyocsik, J. Electrochem. Soc. 114, p. 176 (1967).
91. R. D. Hiedenreich, U. S. Pat. No. 2,935,781.
92. D. A. Kiewit, J. Electrochem. Soc. 117, p. 944 (1970).
93. R. D. Wales, J. Electrochem. Soc. 110, p. 914 (1963).
94. S. G. Ellis, J. Appl. Phys. 28, p. 1262 (1957).
95. R. D. Wales, J. Electrochem. Soc. 111, p. 478 (1964).
96. J. J. Loferski and P. Rappaport, J. Appl. Phys. 30, p. 1296 (1956).
97. L. Young, Anodic Oxide Films, p. 329, Academic Press, New York, NY (1961).
98. S. Zwerdling and S. Sheff, J. Electrochem. Soc. 107, p. 338 (1960).
99. J. P. McKelvey and R. L. Longini, J. Appl. Phys. 25, p. 634 (1954).
100. J. A. Nemetz and R. E. Tressler, Solid State Tech., p. 209, September, (1983).
101. Q. Hua, J. Rosenberg, J. Ye and E. S. Yang, J. Applied Physics 53, p. 12 (1982).
102. Y. Igarashi, K. Kerumada and Niimi, Japan J. Appl. Phys. 300, (1968).
103. H. J. Stein, J. Electrochem. Soc. 121, p. 1073 (1974).
104. G. D. Bagratishvili, R. B. Dzhanelidze and D. A. Jishiashvili, Phys. Stat. Solidi. 78, p. 115 (1983).
105. J. J. Hantzpergue, Y. Doucet, Y. Pauleate and J. C. Remy, Annal. de Chimie 10, p. 211 (1975).
106. G. D. Bagratishvili, R. B. Dzhanelidze and D. A. Jishiashvili, Phys. Stat. Solidi. 78, p. 301 (1983).
107. H. Nagai and T. Niimi, J. Electrochem. Soc. 115, p. 671, (1968).
108. Y. Pauleau and J. C. Remy, Comptes Rendus, HCb: Des Seances De L'Acadamie Des Sciences 280, p. 1215, (1975).
109. G. D. Bagratishvili, R. B. Dzhanelidze and D. A. Jishiashvili, Proceedings of the International Conference on the Chemistry of Semiconductors Heterojunction Layer Structure, 5, p. 65 (1970).

110. R. Storr, A. N. Wright and C. A. Winkler, Canadian J. Chem. 40, p. 1296 (1962).
111. S. N. Ruddlesden and P. Popper, Acta. Cryst. 11, p. 465 (1958).
112. S. Wild, P. Grieveson and K. H. Jack, Special Ceramics, Vol. 5, P. Popper editor, p. 385 (1972).
113. R. Juza, Angew. Chem. 58, p. 25 (1945).
114. W. C. Johnson and G. H. Ridgley, J. Am. Chem. Soc., 56, p. 2395 (1934).
115. W. C. Johnson, J. Am. Chem. Soc. 52, p. 5161, (1930).
116. G. H. Morey and W. C. Johnson, J. Phys. Chem. Soc. 54, p. 3603 (1932).
117. G. H. Morey, Presented at Amer. Chem. Soc. Meeting, (April 1931).
118. R. B. Guthrie and F. L. Riley, Proc. Brit. Ceram. Soc. 22, p. 9275 (1973).
119. B. J. Dalgleish, H. M. Jennings and P. L. Pratt, Science of Ceramics 10, P. Popper editor, British Ceramic Research Association (1982).
120. O. Gregory and M. H. Richman, Metallography 15, p. 157 (1982).
121. P. Grieveson, K. H. Jack and S. Wild, Special Ceramics, Vol. 4, edited by P. Popper, p. 237, British Ceramic Research Association (1968).
122. S. Wild, P. Grieveson and K. H. Jack, Special Ceramics, Vol. 5, edited by P. Popper, British Ceramic Research Association (1972).
123. I. Kohatsu and J. W. McCauley, Mater. Res. Bull. 9, p. 917, (1974).
124. K. Kato, Z. Inoue, K. Kijima, I. Kawada, H. Tanaka and T. Yamane, J. Amer. Ceram. Soc. 58, p. 90 (1975).
125. S. Wild, H. Elliott and D. P. Thompson, J. of Mat's. Soc. 13, p. 1769 (1978).
126. J. C. Remy and Y. Pauleau, J. Nucl. Inorg. Chem. 15, p. 2308 (1976).
127. Y. I. Ukhanov, Y. N. Volgin and F. F. Grekov, Fiz. Elektron. Nauchn Dokl Gertsenovskie Chteniva, 3, p. 45 (1974).
128. R. S. Bradley, D. C. Munro and M. Whitfield, J. Nucl. Inorg. Chem. 28, p. 1803 (1966).
129. O. Kubaschewski and E. H. Evans, Metallurgical Thermochemistry, Oxford, Pergammon Press, (1951).
130. B. J. Dalgleish, H. M. Jennings and P. L. Pratt, Special Ceramics, Vol. 7, D. Popper Editor, British Ceramic Research Association, p. 85, (1981).
131. A. Atkinson, A. J. Moulson and E. W. Roberts, J. Amer. Ceram. Soc. 59, p. 285 (1976).
132. J. G. Gilles, Rev. Hautes. Temp. Refract. 237, p. 62 (1965).
133. J. C. Labbe, F. Duchey and M. Billy, C.R. Hebd. Services Acad., Sci. Serv. C, 273, p. 1750 (1971).
134. P. Grieveson, Nitrogen Ceramics, ed. F. L. Riley, Noordhoff, Leyden, p. 153 (1977).
135. H. M. Jennings, J. Mat. Sci. 18, p. 951, (1983).
136. R. B. Bird, W. E. Stewart and E. N. Lightfoot, Transport Phenomena, Wiley Press, (1960).
137. R. E. Treybal, Mass Transfer Operations, McGraw-Hill (1980).

138. D. Canipos-Loriz and F. L. Riley, J. Mater. Sci. <u>11</u>, p. 195, (1976).
139. H. M. Jennings, and M. H. Richman, J. Mat. Sci. <u>11</u>, p. 2087 (1976).
140. J. W. Evans and S. K. Chatterji, J. Phys. Chem. <u>62</u>, p. 1064 (1958).
141. S. P. Muracka, C. C. Chang and A. C. Adams, J. Electrochem. Soc. <u>126</u>, p. 996 (1979).
142. T. Ito, T. Nozaki and H. Ishikawa, J. Electrochem. Soc. <u>127</u>, p. 2053 (1980).
143. P. Longland and A. J. Moulson, J. Mater. Science <u>13</u>, p. 2279 (1978).
144. A. J. Moulson, Review: J. Mater. Science <u>14</u>, p. 1017 (1979).
145. D. S. Thompson and P. L. Pratt, <u>Science of Ceramics 3</u>, editor G. H. Stewart (1967).
146. Y. Hayafiyi and K. Kayiwara, J. Electrochem. Soc. <u>129</u>, p. 2102 (1982).
147. J. A. Hedvall, <u>Reactions Fahigkeit Fester</u>, J. A. Bacth, Lupzig, 1938.
148. G. D. Bagratishvili, R. B. Dzhaneldye, N. I. Kurdiani, Y. I. Paskintsev, O. V. Saksaganski and V. A. Skoukov, Thin Solid Films <u>56</u>, p. 209 (1979).
149. T. Yashiro, J. Electrochem. Soc. <u>129</u>, p. 280 (1982).

RECEIVED December 26, 1984

Vapor-Phase Epitaxy of Group III–V Compound Optoelectronic Devices

G. H. Olsen

Epitaxx, Inc., Princeton, NJ 08540

The major types of VPE growth systems throughout the world are reviewed. Growth parameters and equipment sketches are included. Limitations and solutions to the "preheat problem" are discussed. Multibarrel reactors, which allow multilayer heteroepitaxial layers to be grown without interruption of crystal growth are also described. Novel growth phenomena such as "non-planar" VPE lasers, lateral growth over dielectric layers, growth on (311) and (511) InP substrates and combined VPE/LPE growth are discussed. In-situ laser diagnostic probing and device results -- including the use of VPE for visible lasers -- are also reviewed.

The vapor phase epitaxy (VPE) of III-V compounds -- whereby solid epitaxial layers are deposited by passing chemical vapors over a substrate -- is of great use in the optoelectronics area. Commercial devices which are synthesized by this process include GaAsP LEDs, GaAs photocathodes, 1.06 μm InGaAs LEDs and lasers, 1.3 μm InGaAsP LEDs, 1.3 μm InGaAsP cw lasers and 0.9-1.7 μm InGaAs PIN detectors (1) (another important class of VPE grown components (2) -- microwave devices -- will not be discussed in this review). Annual sales of optoelectronic devices made by the VPE process undoubtedly run into the tens of millions of dollars.

The two methods for the VPE growth of III-V compounds have been the so-called "chloride" method, whereby a group V chloride (e.g., $AsCl_3$) passes over a metal to form the group III chloride (e.g. GaCl), whereas in the so-called "hydride" technique, the group V element is introduced as a hydride (e.g., AsH_3) and HCl gas is passed over a metal to form the group III chloride. Strictly speaking, the term "halide" should be used rather than "chloride," since I and Br have also been used as transport agents. The main advantage of the chloride system is that it has produced very low ($<1 \times 10^{13}$ cm^{-2}) undoped background impurity concentrations (3) in GaAs. Its main disadvantage is that gaseous $AsCl_3$ is introduced by heating a liquid, and therefore its concentration will vary exponentially with temperature. The hydride system, on the other had, has the advantage that all input reactants to the system are gaseous and can be carefully

controlled in a linear manner. Thus, crystal compositions can be
more carefully controlled with the hydride system, and this is the
major reason for its use -- especially with optoelectronic devices.

VPE Reactor Design

The VPE process in its present form is largely an evolution of
the method first demonstrated by Tietjen and Amick (4) for the growth
of Ga(As,P) alloys using the hydrides arsine (AsH_3) and phosphine
(PH_3). Therefore, the present-day RCA hydride VPE system for the
growth of (Ga,In(As,P) alloys will be described in some detail (5,6).
Other systems will then be described with differences pointed out.
Reports on the hydride/chloride VPE growth of III-V alloys for
optoelectronics include at least Beuchet and coworkers (7) at
Thomson-CSF (France), Enda and coworkers (8) at Nippon Telegraph and
Telephone (NTT) (Japan), Enstrom and coworkers (9) at RCA (USA),
Hyder and coworkers (10) at Varian (USA), Johnston and coworkers (11)
at Bell Laboratories (USA), Kanbe and coworkers (12) at NTT, Mizutani
and coworkers (13) at Nippon Electric Co. (NEC) (Japan), Nagai and
coworkers (14) at NTT, Olsen and coworkers (15) at RCA, Saxena and
coworkers (16) at Varian, Seki and coworkers (17) at Tokyo University
of Agriculture and Technology (Japan), Sugiyama and coworkers (18) at
NTT, Susa and coworkers (19) at NTT, Vohl (20) at Lincoln Laboratories (USA), Zinkiewicz and coworkers (21) at the University of
Illinois (USA) and Jones (22) at Colorado State University. Growth
details for some of these systems are summarized in Table I. Much
early work (23,24) was done on the VPE growth of GaAsP and GaP for
LED applications. Commercial systems for the growth of these materials have been available for some time.

The RCA single-barrel VPE growth system is shown schematically
in Fig. 1. The reactor tube (∿25 mm internal diameter) is made of
quartz except for areas that are not heated to high temperatures
which are made of Pyrex. Heat is provided by the use of "clamshell"
resistance furnaces which surround the tube. The furnaces are left
on at all times (except for disassembly or cleaning) and hydrogen
flows through the tube constantly. Deposition is initiated by
passing HCl gas over the indium and/or gallium metal (which is held
at 850-900°C) in order to form metal chlorides. The area of indium
metal held in quartz boats is considerably larger (∿100 cm^2) than
that of gallium metal (∿ 25 cm^2). Arsine and/or phosphine (10% in
H_2) are brought in through a separate tube and then mixed with the
metal chlorides in a mixing zone. P-type doping is accomplished by
heating a zinc bucket in a hydrogen atmosphere in order to obtain
elemental zinc vapor which is then transported by H_2. N-type doping
is accomplished by adding about 100 ppm H_2S gas to the group V line.
All input reactant flows are controlled by electronic mass-flow
controllers. Growth is initiated by inserting a polished substrate
on a rotatable sample holder with a quartz spring which is attached

Table I. Summary of Growth Techniques

	Enda	Sugiyama	Johnston	Susa	Kanbe	Seki	Zinkewitz	Hyder	Olsen	Mizutani	Vohl
λ (μm)	1.38	0.6-1.0	1.5	1.6-1.7	1.6-1.7		1.3-1.7	1.0-1.7	1.3-1.7	1.3	1.3
HCl Source	HCl/AsH_3	$AsCl_3$/AsH_3	HCl/AsH_3	$AsCl_3$/AsH		HCl/AsH_3	HCl/AsH_3	HCl/AsH_3		HCl/AsH_3	PCl_3
% AsH,PH_3	5%	5%			5%		10%		10%		PCl_3
T_{source}	830	830	840	800	800		750-800	850	850	800	750-820
T_{mix}	900	900		825	825		750		850	850	830
T_{grow}	660	650-725	700	730	650-750	650-700	650	690	700	650-700	700
ΔT (°C/cm)	14	10			1			5°C/cm	0.1°C/cm	<0.5°C/cm	
Preheat	As/PH_3 chamber	none	none	sliding quartz boat	AsH_3 atmos	near baffle	preheat zone	sliding quartz	AsH_3/PH_3 preheat		
H_2 (total)		665	5000		500	400	500-900	1100	5000	1800	
Growth Rate	5-8	5-15			6-12	5			~25		1-3 μ/h
Lowest N_D 300				InGaAs 1×10^{15}	8×10^{15}			8×10^{14}	3×10^{15}		1.7×10^{16}
77				1×10^{15}				4×10^{15}			2.7×10^{16}
Highest μ 300				10,000	5500			4,000	7900		3500
77				35,400					19,000		5800
In Area							14 cm^2	16 cm^2	100 cm^2		
Comments		40 mm quartz stoich. Ga	Dieth Zinc computer dieth-tellur 5% HCl 2% HCl		45 mm diam	single flat temp.	magnetic rotator vertical indium buckets	40 mm diam slider boat 3/1 metal rich		-	(110) growth PCl_3 $AsCl_3$ GaAs,InP,InAs

to the end of the quartz rod and inserted into the forechamber which is then flushed out with hydrogen while the input gas flows are being equilibrated. The substrate is sealed off from the growth chamber by a large Pyrex stopcock (substrate entry valve). After the forechamber is flushed out ($\sim$15 min at 2000 cm^3 min of H$_2$), the stopcock is opened and the quartz rod (which is supported by a close-tolerance 'truebore' gas bearing) is pushed in so that the substrate is moved to the preheat zone. This zone has an atmosphere of arsine and/or phosphine, corresponding to the substrate group V constituent, in order to minimize decomposition effects. The substrate is heated to a temperature near the growth temperature and then inserted into the growth zone where deposition takes place. The substrate is rotated ($\sim$10 rpm) during growth in order to smooth out any non-uniformities in temperature or gas flow. The temperature over the substrate is constant to within $\pm$0.1°C. Thickness uniformities of $\pm$5% and compositional uniformities of $\pm$0.1 mol% have been measured with In$_{0.5}$Ga$_{0.5}$P grown on GaAs. Growth is ended by withdrawing the substrate to the forechamber (before altering any of the flows) where it cools to room temperature in a hydrogen ambient. If the substrate is to be removed, the stopcock must first be closed. However, if subsequent layers are to be grown, the substrate is held in the forechamber while the reactant flows are changed and equilibrated (typically 15-30 min). The above process is then repeated. If compositional grading is desired, the control voltage to the appropriate mass-flow controller can be varied either abruptly (in discrete steps) or smoothly in a continuous fashion.

The Varian system (10) (see Fig. 2) is somewhat similar to the RCA system in design although no provisions are made for a separate preheat zone. The substrate is "preheated" by using a quartz "slider-boat" in which the substrate can be left in the growth zone and protected from gas ambients until growth begins by pulling back the slider top. A thermocouple is also embedded in the holder. Total H$_2$ flow quoted is limited to 1100 cm^3 min^{-1}. A two-zone heat-pipe furnace is used for heating.

The University of Illinois system (22) is also similar to the RCA system except that the growth tube is vertical and total H$_2$ flow is 500-900 sccm (standard cm^3 min^{-1}). A novel system of suspended indium boats (see Fig. 3) is used to ensure intimate contact of HCl gas (with the metal sources). A metal surface area of only 14 cm^2 is employed. A magnetic feedthrough is used to lift and rotate the substrate.

The system of Johnston and Strege (11) is a horizontal one which uses 5000 sccm total H$_2$ flow. A sketch is shown in Fig. 4. No preheat zone is provided. Note that 2% and 5% HCl concentrations in H$_2$ are used to generate GaCl and InCl. Diethyl-telluride (n-type) and diethyl-zinc (p-type) were the preferred dopant sources. The apparatus consists of a quartz reaction vessel with four inlet tubes

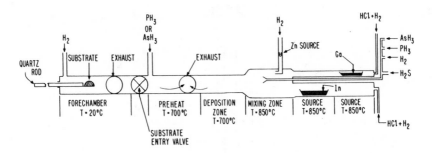

Figure 1.   Sketch of RCA VPE system.

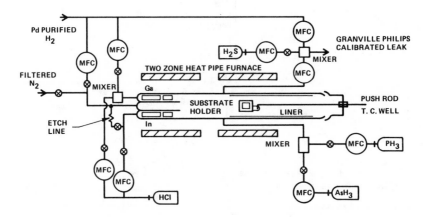

Figure 2.   Varian VPE reactor (courtesy of S. B. Hyder).

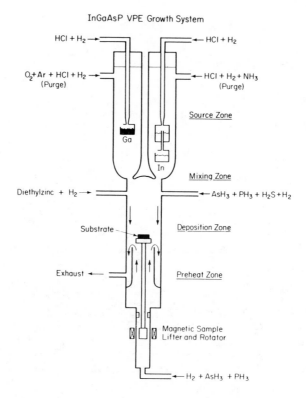

Figure 3.   VPE reactor described by Zinkiewicz (courtesy of L. M. Zinkiewicz).

connected to a single larger growth tube. All gas flows are controlled at the inlet end by electronic mass-flow controllers. AsH_3
and PH_3 flow directly into the growth tube while the HCl is passed
over graphite boats containing liquid indium and liquid gallium.
Dopants are added through the fourth tube. Mixing of the metal
chlorides and the hydrides is delayed by passing the hydride gases
through a nozzle. This reduces wall deposits upstream from the substrate, which may cause changes in growth rate and composition. The
nozzle tip is perforated to enhance active gax mixing just before
flowing over the substrate. The substrate is pushed in from the
exhaust end to a point about 5 cm downstream from the nozzle. The
reactor is mounted in a horizontal furnace and heated to a temperature of 840°C at the source zone and 700°C at the growth zone.

The system of Sugiyama et al (18) is shown in Fig. 5. $AsCl_3$ is
used as an HCl source. A residence time of over 4 s is claimed
between the metals and HCl gas. No provision for p-type doping is
specified.

The system of Enda (8) employs a sealed preheat chamber in which
the substrate temperature can be brought up to growth temperature in
a group V hydride ambient (see Fig. 6).

The system of Susa et al (19), primarily for InP and InGaAs
growth, uses $AsCl_3$ as an HCl source and also uses a sliding quartz
boat to protect substrates from decomposition. Improved crystalline
quality was attributed to this boat.

Vohl (20) has grown alloys by the trichloride method which uses
PCl_3 and $AsCl_3$ as sources for P and As. Three compound sources, InP,
InAs, and GaAs -- each in a separate tube -- are reacted with a gas
mixture containing the same group V element. Each reaction is
independently controlled. The deposition of the alloys is regarded
as the deposition of a mixture of GaAs, InP, and InAs. The alloy
composition is obtained by adjusting the relative amounts of the
three compounds that are deposited on the substrate. A sketch of the
furnace is shown in Fig. 7.

Limitations Imposed by the Preheat Process

One drawback of the single-barrel VPE process is the preheat
cycle. Here, substrate wafers are heated up to 650-700°C, prior to
growth, usually without deposition or etching taking place. This
temperature range equals or exceeds the congruent evaporation temperature (25) for most III-V compounds, and some preferential evaporation of the group V element is bound to take place. InP is particularly susceptible to this phenomenon, since its decomposition temperature is only around 400°C. The damage can be reduced by preheating

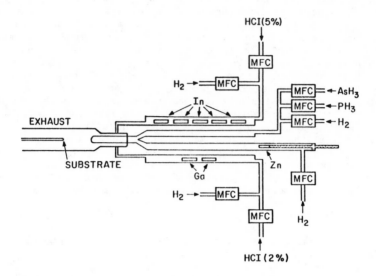

Figure 4.   VPE reactor of Johnston and Strege (courtesy of
W. D. Johnston, Jr.).

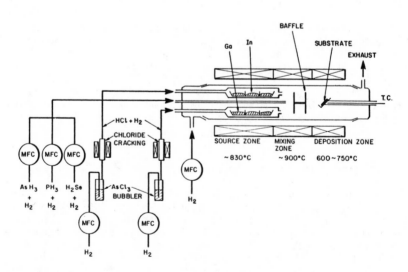

Figure 5.   VPE reactor of Sugiyama et al. (courtesy of
K. Sugiyama).

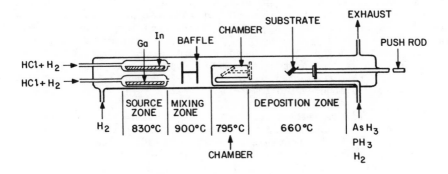

Figure 6.   VPE reactor of Enda (courtesy of H. Enda).

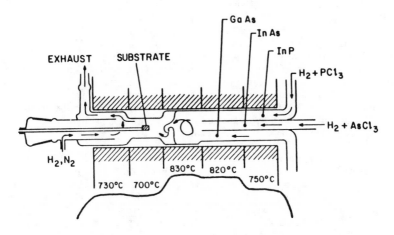

Figure 7.   VPE reactor of Vohl (courtesy of P. Vohl).

the substrate in the appropriate group V hydride (1) (i.e., AsH_3 or PH_3).

Although the surface morphology of a preheated layer is obviously degraded by the preheat decomposition, the microstructure of the material also suffers. A high density of defect loops has been observed, (1) via transmission electron microscopy, near the active regions of InGaP/GaAsP lasers. The dislocation loops can be eliminated by using high flows of the group V carrier-gas and lower temperatures during preheat. However, an excess point-defect concentration may always be present to some extent, and should be viewed as a potential failure mechanism in any III-V compound light-emitting device prepared at elevated temperatures.

Preheat decomposition can be reduced with the use of the "slider-boat" substrate holder whereby a quartz "slider" is used to shield the substrate from gaseous ambients (10). The NTT version (19) allows group V gases to be trapped along with the substrate. However, microscopic decomposition still probably occurs.

Multibarrel VPE Reactors

The preheat problem, discussed in the last section, can be solved by eliminating the need for it. The use of multibarrel reactors allows successive heteroepitaxial layers to be grown by moving the susbstrate from one growth chamber to another without interrupting growth. A dual-growth-chamber GaAs VPE reactor was proposed by Watanabe et al (26) in 1977 to produce abrupt changes in doping profile. This system was upgraded into a GaInAsP dual-growth-chamber reactor by Mizutani et al (13). Other types of multibarrel reactors were independently conceived to grow (Ga,In)(As,P) by Olsen and Zamerowski (15) (the "double-barrel" reactor) and by Beuchet et al (7) (the four-barrel "multichamber" reactor). The three types of reactors are shown in Figs. 8, 9, and 10.

The "dual-growth-chamber" reactor of Mizutani (13) (Fig. 8) consists of one horizontal chamber to grow InP and another parallel chamber to grow alloys of (Ga,In)(As,P). The substrate is supported by a rod mounted on a flexible rubber bellows so that transfer between chambers is accomplished by withdrawing and shifting the rod. Transfer can be effected within 2 s and heterointerface transition widths less than 50-60 Å are claimed. Although (in the form shown) gallium-bearing alloys can only be grown in one of the chambers, and no provisions have been made for p-type doping, high-quality 1.3 μm cw lasers have been synthesized in the system (Cd was diffused in after growth for p-contact).

The "multichamber" reactor of Beuchet et al (7) (Fig. 9) consists of four parallel tubes, each with a specific function. The

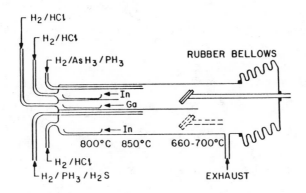

Figure 8. "Dual-growth-chamber" VPE reactor of Mizutani (courtesy of T. Mizutani).

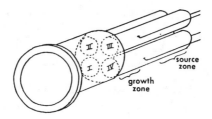

Figure 9. "Multichamber" VPE reactor of Beuchet (courtesy of G. Beuchet).

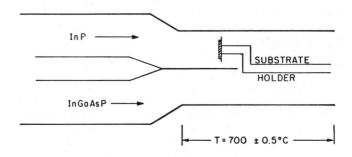

Figure 10. RCA "double-barrel" VPE reactor.

system has been automated so that interchamber transfer can be effected within 1 s. The purpose of each chamber is as follows:

first: preheat and substrate etch.
second: growth of n-InP
third: growth of n-GaInAs
fourth: growth of p-InP

Metallic zinc is used for p-type doping while H_2S is the n-type dopant.

The "double-barrel" VPE reactor (15) designed at RCA is shown in Figs. 10. The concept involves the use of two conventional VPE systems placed in parallel and feeding into a single-growth chamber. With this construction, different gases can be run through each tube, so that double-heterostructure lasers (e.g., InP/InGaAsP/InP/InGaAsP) can be prepared by simply switching the substrate from one tube to the other, thus removing the need for preheat cycles, which may limit device performance since they can introduce interfacial defects. It is important to note here that <u>crystal growth is not interrupted during the the switchover from one layer to another.</u> The substrate can either be held just in front of each tube, or inserted into the tube if any mixing problems occur. Conversely, by holding the substrate at the end of the quartz plate and rotating it, extremely thin multilayers can be grown. It is estimated that by rotating the substrate at 200-300 rpm under normal VPE growth conditions, single atomic layers of each material could be deposited.

Figure 11 contains an SEM photo of a stained angle-lapped section from a 50-layer InAsP/InP multilayer structure -- each with a 200 Å thickness.

Although exciting fundamental studies may be performed with structures prepared in the double-barrel reactor, its main practical advantage is a saving in growth time and materials, as well as the possible attainment of highly-superior interfacial properties. Growth of a typical laser structure presently takes about 2 hours and involves four preheat cycles, including one at each of the two cavity interfaces. InP is particularly susceptible to decomposition effects at elevated temperatures. With the new system, the total growth time is reduced to 30 min, and all preheat cycles are eliminated (except for the original one, which can be positioned 1-10 μm below the laser cavity). The reduced growth time would enable 15 or more laser wafers to be grown in a single day. The elimination of the preheat cycles is expected to yield cleaner interfaces, which should provide lasers with lower threshold current densities, higher efficiencies, and better reliability. Improved interfaces should similarly enhance the performance of avalanche photodiodes, as discussed in the previous section. These advantages are highlighted in Table II.

Table II. Advantages of Double-Barrel VPE Reactor

• Cleaner Interfaces: Crystal growth is not interrupted when growing multiple (heteroepitaxial) layers.

• Rapid Device Growth: A complete four-layer InGaAsP/InP/InGaAsP/InP 1.3 µm cw laser structure can be grown in about 30 minutes. Fifteen or more wafers of this type could be grown in one day.

• Very-thin Layer Growth: Multiple layers ($\leq$200 Å) can be grown, thus allowing "quantum-well" type structures to be fabricated. By rotating a substrate at $\sim$250 rpm in front of the two barrels, layers on the order of 10 Å might be grown.

• Cheaper: A factor six savings in material (gases, metals, etc.) costs results from the reduced growth time for devices.

Novel Growth Phenomena

a. Anisotropic VPE Growth

The growth rate of VPE layers depends markedly on substrate orientation. Variations up to 60:1 have been observed (20) under identical growth conditions. Figure 11 contains a photograph of a stained cleavage edge from a hydride VPE InP grown over etched "dovetails" in an InP substrate. Note the suppression of growth on the (111) facets of the dovetail and the tendency for flat planar growth on the (100) substrate surface. The "thick" growth shown in Fig. 11 indicates that eventually the dovetail fills in ("mirroring" the dovetail in the epitaxial layer) and a flat planar layer is observed everywhere. This behavior contrasts with LPE growth over channeled substrates (27) whereby rounding of the channel and non-planar growth is usually observed.

Non-planar growth of this type can also be achieved with VPE by growing over mesas which have been chemically "rounded-off" after formation. Figure 12 contains optical and SEM photos of such structures. Further details on these structures may be found elsewhere (28). Non-planar lasers of this type have been fabricated (29) with cw lasing threshold currents as low as 45 mA and cw power outputs beyond 70 mW.

Figure 11. SEM photo of a stained cleavage edge from an InGaAs/InP multilayer structure grown in the double-barrel VPE reactor. 60Å layer thicknesses are evident.

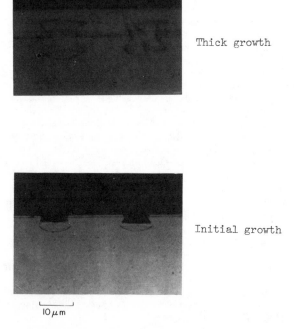

Figure 12. Optical photograph of a stained cleavage edge of VPE InP grown over etched "dovetails" along [011] direction in (100) InP.

Workers at Lincoln Labs (20,30) have reported another unusual anisotropic growth effect with VPE: lateral overgrowth. Thin sheet-like epitaxial layers have been grown over SiO_2 masked III-V compound substrates by opening up narrow channels in the SiO_2 and then growing a VPE layer. With proper choice of substrate orientation (typically {110} or {100}) and channel direction, single crystal growth initiates in the channel and the spreads out laterally over the glass at a much faster rate than the growth perpendicular to the substrate. This opens up the possibility of "reusable" substrates as lateral overgrowths of this type have actually been broken off from their substrates. Lateral overgrowth has been demonstrated with GaAs (30) and InP (20).

b. VPE Growth on (311) and (511) Substrates

Olsen et al (31) have recently reported improved VPE growth (relative to (100)) of InP and InGaAs on (311) and (511) InP substrates. Even more recently (32), cw lasing performance was achieved with 1.3 μm InGaAsP lasers grown on these orientations. Lasing threshold current densities with (311) devices were actually lower than with (100) devices and photoluminescence from p-InGaAsP "cap" layers were 20-30 times higher on (311) and (511) devices. However, good crystal growth and device results have been observed numerous times for non-standard substrate orientations. The point here is that while most III-V compound devices have been prepared on standard (100) and (111) substrates, non-standard substrate orientations might well be a fruitful area of study both from a fundamental and a technological point of view. See Figure 13.

c. Combined Vapor and Liquid Phase Epitaxy

Difficulties have been reported with the LPE growth of InP upon InGaAsP alloys which are lattice-matched to InP, and have bandgap wavelengths beyond 1.4 μm, due to dissolution of the InGaAsP (including InGaAs) solid substrate layer in the InP melt. Although several "tricks" (e.g., rapid cooling; low growth temperatures) have been developed to circumvent this problem, another cure is the VPE growth of InP upon other layers grown by LPE. Workers at NTT have succeeded in fabricating both InGaAs avalanche photodiodes (33) and 1.5 μm buried heterostructure lasers (34) with combined VPE and liquid phase epitaxy (LPE).

VPE has also been used (35) to grow highly Zn-doped p-InGaAsP "contact" layers upon LPE grown 1.3 μm InGaAsP/InP laser structures.

A novel device application (36) involves a mass transport phenomenon (37). A buried "terrace"-type 1.3 μm InGaAsP laser was grown (36) by H_2 "vapor" transport of InP from an InP wafer onto an etched LPE laser. Lasing threshold currents as low as 9.5 mA were reported (36).

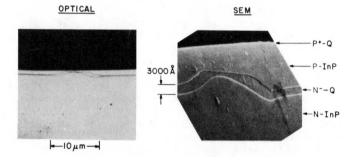

VPE GROWTH OF MODE-CONFINING STRUCTURE
ON (11O) InP SUBSTRATE

Figure 13.   Optical and SEM photos from a stained cleavage edge
of a "non-planar" 1.3μm InGaAsP cw laser.

Laser Diagnostic Probes for VPE Growth

Donnelly and Karlicek (38) have developed an interesting method for in-situ monitoring of reactant species during the hydride VPE growth of InP and InGaAsP. Laser induced fluorescence of the gas-phase species was used to determine vapor concentrations under actual crystal growth conditions in a VPE reactor. In addition to deter-mining the concentration of such species as P_2, PH_3, InCl as a function of temperature and other growth variables, several other pieces of experimental information evolved from the study. GaCl concentration was found to vary linearly with HCl flow while InCl concentration had a non-linear dependence. PH_3 decomposition was found to be incomplete during deposition. The delay time between initial HCl turn-on and final steady-state InCl concentration was about 1 min (with 500 ccpm total flow. Similarly the lag time between AsH_3 turn-off and disappearance of As-species was several minutes. This was attributed to transport of As from wall deposits.

Laser diagnostic techniques have exciting possibilities. In-situ monitoring could identify vapor concentrations wich give best surface morphologies and device results allow corrections for de-pleted metals, wall deposits etc. and improve the reproducibility of VPE growth in general (38).

Device Results

Perhaps the best recommendation for the use of the VPE technique is the impressive list of device results which have been obtained from structures grown with this technique. Among these are (see ref. 15):

1. Transmission secondary electron (TSE) devices made from InGaP/GaAs structures which exhibit gains as high as 540 at primary energies of 20 kV.

2. Negative electron affinity (NEA) GaAs reflection photo-cathodes which exhibit sensitivities as high as 2150 μA/lm.

3. NEA GaAs/InGaP/GaP transmission photocathodes which exhibit quantum efficiencies as high as 15% at 8500 Å.

4. GaAs solar-cells covered with thin (~200 Å) InGaP layers which exhibit short-circuit current densities as high as 22 mA/cm^2, open circuit voltages as high as 0.96 V, and conver-sion efficiencies at AMI of up to 14%.

5. Room temperature double-heterojunction cw lasers which emit light at 7000 Å (GaAsP/InGaP), 8800 Å (GaAs/InGaP), 1.06 μm (InGaAs/InGaP), 1.25 μm and 1.4 μm (InGaAsP/InP), which have lasing threshold current densities of 1000-3400 A/cm^2 and differential quantum efficiencies of 50% or more.

6. InGaAsP/InP 1.3 μm edge-emitting LEDs which exhibit power efficiencies over 1%, and can couple over 100 μW into an optical fiber.

7. Efficient red and green GaAsP and GaP LEDs (39).

Johnston and Strege (11) have made 1.52 μm GaInAsP/InP VPE lasers with pulsed threshold current densities below 1200 A cm^{-2}. Mizutani et al (13) have demonstrated cw lasing action near 1.3 μm with VPE GaInAsP/InP lasers grown with their new "dual-growth-chamber" reactor. Susa et al (40) and Ando et al (41) have demonstrated impressive detector results with 1.0-1.7 μm $Ga_{0.47}In_{0.53}As$/ InP VPE structures including a SAM (separate absorption and multiplication) device whereby light is absorbed in the low-bandgap $Ga_{0.47}In_{0.53}As$ ($E_g \sim 0.74$ eV) but multiplied via avalanche gain the high-bandgap InP ($E_g \sim 1.35$ eV) region.

Future Directions

As long-wavelength (1.0-1.7 μm) GaInAsP electro-optical devices become more widely used, motivated by (a) low fiber absorption and dispersion; (b) high transmission through water and smoke; and (c) greatly enhanced eye safety at wavelengths greater than 1.4 μm, the VPE process should become increasingly widespread. Vapor-phase epitaxy offers its usual advantages of smooth, uniform surfaces, uniformly low background doping density, and economy of scale for large-area devices (e.g., large detectors). However, a particular advantage exists for the VPE growth of devices for wavelengths above 1.4 μm compared to liquid-phase epitaxy (LPE). The LPE growth of InP onto GaInAsP alloys which have $\lambda_g > 1.4$ μm is difficult because of dissolution of the alloy by the InP melt. Although this problem can be circumvented by special growth techniques, it does limit the flexibility of the LPE technique since the usual growth techniques cannot be used here. No such problems exist for VPE, and multilayer heterostructures with bandgap wavelengths in excess of 2.0 μm can be fabricated using the same techniques as for 1.3 μm devices. This would include emitters and detectors in the 2-3.5 μm range (using InGaAs/InAsP) and the 3-6 μm range (using InGaSb/InAsSb). As long-wavelength GaInAsP lasers, LEDs, and photodetectors begin to replace GaAs emitters and Si detectors for low-loss fiber applications and as new applications arise in the 1.0-1.7 μm regime, look for the VPE process to play an increasingly important role in the technology of electronic devices.

Another important application concerns visible lasers. VPE can be used to grow InGaP/GaAsP and InGaP/InGaAsP heterostructures for lasers which emit light near 7000 Å. CW operation at 10°C was reported for 7050 Å InGaP/GaAsP lasers by Kressel (42) et al. Usui et al (43) have reported cw room temperature operation near 7600 Å with VPE InGaAsP/InGaP. (They also reported reduced wall deposits with the addition of small oxygen concentrations). As interest in visible lasers increases, driven by optical recording/playback and audio disc applications, VPE technology will undoubtedly become more widespread in optoelectronic applications.

Acknowledgments

This paper was originally presented at the Spring 1983 Meeting of The Electrochemical Society, Inc. held in San Francisco, Calif. Copyright 1983, The Electrochemical Society.

Literature Cited

1. G. H. Olsen, GaInAsP Alloy Semiconductors, pp. 1-41, ed. T. P. Pearsall, J. Wiley, London, 1982.
2. RCA Review, 2, pp. 499-782 (1981) (Special Issue on Microwave Technology).
3. R. D. Fairman, M. Omuri and F. B. Fauk, Gallium Arsenide and Related Compounds (St. Louis), 1976, Conf. Ser. No. 33b, Institute of Physics, London, 1977, pp. 45-54.
4. J. J. Tietjen and J. A. Amick, J. Electrochem. Soc., 113, 724-728, 1966.
5. G. H. Olsen and T. J. Zamerowski, IEEE J. Quantum Electron., QE-17, pp. 128-138, 1981.
6. G. H. Olsen and M. Ettenberg, Crystal Growth: Theory and Technique, Vol. II, ed. C. H. L. Goodman, New York, Plenum Press, 1978, pp. 1-56.
7. G. Beuchet, M. Bonnet and J. P. Duchemin, Proc. 1980 NATO Conf. on InP, Rome Air Development Center Tech. Memo, RADC-TM-80-07, Hanscom Air Force Base, MA, 1980, p. 303.
8. H. Enda, Jpn. J. Appl. Phys., 18, pp. 2167-2168, 1979.
9. R. E. Enstrom, D. Richman, M. S. Abrahams, J. R. Appert, D. G. Fisher, A. H. Sommer and B. F. Williams, 3rd Int. Symp. on GaAs and Related Compounds, Institute of Physics, London, 1970, pp. 30-40.
10. S. B. Hyder, R. R. Saxena and C. C. Cooper, Appl. Phys. Lett., 34, pp. 584-586, 1979.
11. W. D. Johnston, Jr., and K. E. Strege, 38th Annual IEEE Device Research Conf. Abstracts, Cornell University, Vol. IVB-3, June 1980.
12. H. Kanbe, Y. Yamauchi and N. Susa, Appl. Phys. Lett., 35, pp. 603-605, 1979.
13. T. Mizutani, M. Yoshida, A. Usai, H. Watanabe, T. Yuasa and I. Hayashi, Jpn. J. Appl. Phys., 19, L113-L116, 1980.
14. H. Nagai, J. Electrochem. Soc., 126, pp. 1400-1403, 1979.
15. G. H. Olsen and T. J. Zamerowski, Progress in Crystal Growth and Characterization, Vol. II, ed. B. R. Pamplin, London, Pergamon, 1979 pp. 309-375.

16. R. R. Saxena, S. B. Hyder, P. E. Gregory and J. S. Escher, J.
     Crystal Growth, 50, pp. 481-484, 1980.
17. H. Seki, A. Koukitu and M. Matsumura, J. Crystal Growth, 54, pp.
     615-617, 1981.
18. K. Sugiyama, H. Kojima, H. Enda and M. Shibata, Jpn. J. Appl.
     Phys., 16, pp. 2197-2203, 1977.
19. N. Susa, Y. Yamauchi, H. Ando and H. Kanbe, Jpn. J. Appl. Phys.,
     19, L17-L20, 1980.
20. P. Vohl, J. Crystal Growth, 54, pp. 101-108, 1981.
21. L. M. Zinkiewicz, PhD Thesis, Univ. of Illinois (1981).
22. K. Jones, J. Crystal Growth (to be published).
23. R. A. Burmeister, Jr., G. P. Pighini and P. E. Greene, Trans.
     Met. Soc. AIME, 245, 587-591 (1969).
24. W. O. Groves, A. H. Herzog and M. G. Crawford, Appl. Phys.
     Lett., 19, pp. 184-186 (1971).
25. B. Goldstein, Semiannual Report #2, Contract No. DAAK02-74-C-
     0081, Night Vision Lab, Ft. Belvoir, Va. (1975).
26. H. Watanabe, M. Yoshida and Y. Seki, 151st Electrochem. Soc.
     Ext. Abst., pp. 255-256 (May 1977).
27. D. Botez, IEE Proc., 129, pp. 237-251 (1982).
28. T. Zamerowski and G. H. Olsen, Late News Paper (May 1983 Electro-
     chemical Society Meeting, San Francisco).
29. G. H. Olsen, T. J. Zamerowski and N. J. DiGiuseppe, 1982 IEDM
     Extended Abstracts (IEEE Press, New York, 1982).
30. F. J. Leonberger, C. O. Boyler, R. W. McClelland and L.
     Melngailis, Proc. Topical Mtg. on Integrated and Guided Wave
     Optics, Incline Village, Nevada (1980).
31. G. H. Olsen, T. J. Zamerowski and F. Z. Hawrylo, J. Crystal
     Growth, 59, pp. 654-658 (1982).
32. G. H. Olsen, T. J. Zamerowski and N. J. DiGiuseppe, 8th IEEE
     Int. Semiconductor Laser Conf. Abstracts (Ottawa, 1982).
33. N. Susa and Y. Yamauchi, J. Crystal Growth, 51, pp. 518-524
     (1981).
34. O. Mikami, H. Nakagome, Y. Yamauchi and H. Kanbe, Electron.
     Lett., 18, pp. 237-239 (1982).
35. F. Z. Hawrylo and G. H. Olsen, U. S. Patent No. 4,355,396.
36. T. R. Chen, L. C. Chiu, K. L. Yu, U. Koren, A. Hasson, S.
     Margalit and A. Yariv, Appl. Phys. Lett, 41, pp. 1115-1117
     (1982).
37. Z. L. Liau and J. N. Walpole, Appl. Phys. Lett., 40, pp. 568-569
     (1982).
38. V. M. Donnelly and R. F. Karlicek, J. Appl. Phys., 53, pp.
     6399-6407 (1982).
39. C. J. Nuese, H. Kressel and I. Ladany, IEEE Spectrum, 9, 28-38
     (1972).
40. N. Susa, Y. Yamauchi and H. Kanbe, Jpn. J. Appl. Phys., 20, pp.
     L253-L256 (1981).
41. H. Ando, N. Susa and H. Kanbe, Jpn. J. Appl. Phys., 20, pp.
     L197-L199 (1981).
42. H. Kressel, G. H. Olsen and C. J. Nuese, Appl. Phys. Lett., 30,
     pp. 249-252 (1977).
43. A. Usui, Y. Matsumoto, T. Inoshita, T. Mizutani and H. Watanabe,
     GaAs and Rel. Comp., Inst. Phys. Conf. Ser. 63, London, pp.
     137-142, (1981).

RECEIVED March 1, 1985

Advanced Device Isolation for Very Large Scale Integration

H. B. Pogge

East Fishkill Facility, IBM General Technology Division, Hopewell Junction, NY 12533

Device-isolation concepts and methods have changed
over the years, each new approach typically requiring
various elements of process development. The
ultimate device isolation is one in which the devices
can be separated from each other by a dielectric
region that reaches down past the active device areas
and occupies minimal real estate. The device
elements in such a case could all be butted against
the isolation region. This process, in principle,
becomes lithography limited and is extendible with
future lithography improvements. Device scaling is,
therefore, assured. It also permits some process
simplifications in comparison to current process
technologies, such as fewer masks, elimination of
lateral autodoping and Si_3N_4 films, as well as
improved surface planarities. The process
technologies required to achieve this form of device
isolation (generally termed "trench isolation")
relate mainly to reactive ion etching (RIE) and
chemical vapor deposition (CVD) techniques. In both
instances, extensions of current process capabilities
and process understanding had to be developed in
order to accommodate the various isolation
requirements dictated by different device design
configurations. Elements of these processes will be
discussed, along with examples of associated process
difficulties and accomplishments.

The semiconductor device fabrication technology has experienced,
over a relatively short period of time, a rapid transition from the
original discrete transistors (1950s), to integrated circuits
(1960s), and these were followed by large scale integration (LSI)
of devices and circuits in the 1970s. Although this technology
evolution is firmly rooted, on the one hand, to a broad base of
device fabrication improvements and expanded process capabilities,

it has, on the other hand, also been closely associated with some very timely implementations of a number of key innovative processes.

Improved semiconductor devices and circuits tend to be achieved primarily through advances in either device density, performance or functionality. At times, all three elements are realized simultaneously. The improvements can be accomplished in several ways. One approach is to design new circuit families, which very often create enhanced device functionalities and performances. In contrast to this, device density gains typically demand a much greater investment of process efforts in the form of a relentless pursuit of better tolerance control of all the various process and tool parameters. It may, however, also be coupled with the introduction of new process methods and/or tools with enhanced process capabilities.

As an example of the latter point, several process and tool innovations have contributed to both the generation of smaller device patterns and the ability to maintain the integrity of those dimensions through the subsequent process steps. Thus, improved exposure tools (e.g., optical step and repeat, E-Beam) and masking materials (resist) allow considerably smaller pattern definitions than was previously possible. Once generated, these patterns can now be maintained with improved pattern etching techniques (reactive ion or plasma etching).

Furthermore, the change-over from diffusion processes to ion implantations helps in minimizing the lateral movements of doped regions, and thereby retains for the most part the originally generated device pattern dimension.

More dramatic impacts on device density and functionality have occurred from time to time with the introduction of totally different processes or unique process concepts. Simultaneously, they have frequently brought along other benefits, such as overall device process simplifications or expanded possibilities for new devices and circuits.

Over the next few years much attention within the semiconductor process technology will be focused on the fabrication of devices which reach beyond the densities of the LSI chips. To assure this successful migration towards VLSI (very large scale integration), several of the key process areas require major improvements over the currently practiced state-of-the-art. The enhancements must come in several areas: shallower transistors, finer pattern dimensions, enhanced transistor contacts, advanced device interconnections, and improved device isolation structures. Some of the technologies have already evidenced good progress, such as the polycrystalline base and emitter contacts (1,2) and improved interconnecting metallurgy schemes (3). The device isolation methods have, however, not been effectively upgraded since about 1970. In its present form, it alone takes up considerable "real estate" on any chip layout and therefore drastically restrains the desired device density.

From a device designer's point of view, the ideal isolation structure should feature two basic characteristics: a) narrow dimensions to retain those generated by the lithography process, and b) a geometric configuration, which allows the butting-up of all the device and circuit elements against it. The isolation region would have to reach down past all the active device regions, including the buried subcollector underneath the epitaxial layer in the case of bipolar devices. Past and current device isolation approaches do not effectively meet these requirements. In fact, they have very specific limitations towards being able to meet the narrowness and depth requirements due to inherent process limitations.

This paper will discuss a new approach for device isolation, well-suited for applications of VLSI designs and for meeting all the previously mentioned device design requirements. It is a dramatic departure from the traditional isolation structure fabrication methodology, thereby allowing unique process sequencing not previously available. This process technology was initially discussed by the author in several presentations (4-8) under the title of "deep groove isolation." The more generic term "trench isolation" is as appropriate. The process concept has, after its initial presentation in 1979 (5), found quick and widespread acceptance in the process industry, as evidenced by a rapid proliferation of papers on this topic in 1982-1983 (9), and appears to be considered as the best approach to meet the VLSI device isolation requirements in the future.

Historical Background of Device Isolation

The early semiconductor devices were fabricated as single transistors on a silicon chip. The inherent device isolation was, therefore, achieved by physical separation of chips from the silicon wafer. Within a chip, device isolation between the transistors became necessary with the introduction of intergrated circuits, in which a number of transistors occupied space on the same chip. The initial approach was the use of electrically biased diffused regions, the so-called "junction isolation" (10). The diffusion was made from the top surface down through the epitaxial layer and into the substrate, deep enough to separate adjacent subcollector diffusions of bipolar transistors (see Figure 1). The surface area taken up by this diffusion was quite large. This was partly because of the depth of the diffusion being dictated by the epitaxial film thickness, which in the early devices was generally 6-10 µm. In addition, a large portion of the device area was taken up by the lateral diffusion (6-10 µm) to both sides of the patterned diffusion window. The lateral diffusion could not be avoided. Furthermore, it was found that the P+ (boron) isolation diffusion had to be kept far enough away from the N+ (arsenic, phosphorous) subcollector diffusion, so as not to introduce crystallographic defects within the transistor region.

An improvement in the isolation area requirement was achieved with the replacement of the single diffusion by a double diffusion process (11). In this approach, an initial P+ diffusion was placed into the substrate between the subcollector regions prior to the epitaxial growth. The boron in the isolation diffusion region updiffuses into the epitaxial film during the epitaxial deposition cycle. A subsequent boron diffusion downward into the epitaxial layer was then made to meet up with the bottom diffusion to form a completed isolation region (Figure 1). This reduced the overall diffusion cycle drastically, and therefore the lateral diffusion

component as well. However, the process sequence became more
complex, requiring two diffusions, an extra mask, and the
associated additional processing (see Table I). In addition,
cross-contamination of arsenic and boron autodoping during the
epitaxial deposition became a new parameter to be understood and
controlled.
        Another process modification was introduced in the early 1970s
to improve the device isolation. In this approach, the top P+
diffusion was replaced with a thick ($\sim$1.0 μm) silicon dioxide
region. This method is referred to in the literature by a variety
of names: Recessed oxide isolation, LOCOS, Isoplanar, Planox,
SWAMI, SILO (12-17). The specific process sequence and final
geometric configuration differs somewhat among these processes;
however, the basic function remains the same for all of them. The
P+, N+ diffusion separation remained the same as in the earlier
junction isolation. The improved device density was realized, in
that the oxide isolation allowed the butting-up of the base and
emitter regions against the oxide, although not as effectively as
desired. An additional advantage of the oxide was the reduction of
metal line capacitances, when these lines were placed on top of
large area oxide regions.
        The oxide isolation methodology has found wide acceptance in
the process industry as the preferred device isolation scheme,
despite some major shortcomings. For one, it has evolved as a
still more complicated process, requiring several additional
processes (see Table 1). The oxide isolation region is
characterized by two unique features - the so-called "bird's beak"
and the "bird's head" (18,19). The beak tends to prevent fully
butted junction formation of the base and emitter. Thus, the full
benefits of device density cannot be realized. More importantly,
the bird's beak generation represents a real limitation to achieve
much smaller (narrower) device isolation regions, and therefore,
ultimate device densities.
        The formation of the bird's head and beak comes about due to
the use of the Si_3N_4 masking film required for the selective
oxidation of the silicon. The Si_3N_4 film must be deposited on top
of a thin layer of SiO_2; if not, the direct contact of Si_3N_4 with
the silicon substrate would initiate crystallographic defects in
the substrate during the oxidation cycle. On the other hand, with
the presence of the thin SiO_2 film underneath the Si_3N_4 layer, some
additional oxidation occurs in that thin SiO_2 layer at the
perimeter of the Si_3N_4 window. This results in the lifting of the
Si_3N_4 in that area to form the bird's head and extending it
underneath to generate the bird's beak (see Figure 2). Detailed
analyses of the formation of this geometry and its relationship to
numerous process parameters have been made by a great many
investigators (19,20). This has been followed by extensive efforts
to overcome these basic limitations. Some success has been
reported, but only at the cost of additional process complexities.
The SWAMI (16) and SILO (17) processes are examples of this effort.
The methods attempt to seal off the lateral oxidation underneath
the Si_3N_4 film with the addition of a second Si_3N_4 layer. The
technique requires very careful processing with small process

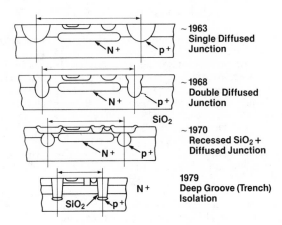

Figure 1   Isolation processes for bipolar devices.

Table I.   Device Process Comparison (for NPN Bipolar)

PROCESS STEP	ISOLATION PROCESS			
	JUNCTION	DOUBLE DIFF	OXIDE/JUNCTION	TRENCH
Substrate Oxidation	X	X	X	X
Subcoll. Pattern Def.	X (Mask A)	X (Mask A)	X (Mask A)	–
Subcoll. Pattern Etch	X	X	X	–
Subcoll. Implant	X	X	X	X
Subcoll. Reoxidation	X	X	X	–
Subisol. Pattern Def.	–	X (Mask B)	X (Mask B)	–
Subisol. Pattern Etch	–	X	X	–
Subisol Implant	–	X	X	–
Oxide Removal	X	X	X	X
Epitaxy	X	X	X	X
Epitaxy Reoxidation	X	X	X	–
CVD Si_3N_4	–	–	X	–
Isol. Pattern Def.	X (Mask B)	X (Mask C)	X (Mask C)	X (Mask A)
Isol. Pattern Etch	X	X	X	X
Isol. Implant	X	X	–	–
Isol. Si Etch (Trench)	–	–	X	X
Isol. Trench Oxidation	–	–	X	X
Isol. Trench Implant	–	–	–	X
Isol. Trench CVD	–	–	–	X
Trench Planarization	–	–	–	X
Excess Material Remov.	–	–	–	X
Total Steps	11	14	16	12
Total Masks	2	3	3	1

windows in order to minimize associated stress concerns and defect
formation (20).
     The oxide isolation has one additional key limitation. The
overall thickness of the oxide is practically limited to about
1.0 μm. Greater thicknesses would require an inordinate amount of
process time (somewhat offset with the use of high pressure
oxidation methods). Furthermore, further thicknesses would also
require additional lateral areas consumed during the oxidation
process. Thus, in the final analysis, this process technology does
not completely have the desireable features for the full
extendibility required for VLSI applications.
     Another direction for device isolation, also introduced in the
early 1970s, was the so-called total dielectric isolation. Several
variations of the process were pursued (21,22). However, wide
acceptance of the process was never realized, primarily because of
the extensive processing requirements. Only in some special
application areas was this technology exercised. And there are no
serious considerations of this process to meet the needs for the
VLSI technology.
     In 1977 Mukai et al. (23) discussed an isolation process that
was based on a V-groove trench, lined with SiO_2, and refilled with
polycrystalline silicon. After removing the excess polycrystalline
silicon from the top of a wafer, the exposed polycrystalline
silicon in the groove was oxidized. No bird's beak is generated,
and the isolation region reached past the epitaxial layer in the
substrate. Selective subcollector regions were used. The main
drawback of this process was that the V-groove required a
reasonable amount of space, with deeper grooves requiring more
lateral space.

The Trench (Deep Groove) Isolation Technology

Due to the inherent limitations of the currently practiced
processes to meet the VLSI device isolation requirements,
development efforts were initiated several years back to search for
an appropriate process technology. One of these efforts has led to
the "trench isolation" technology (5), a process approach
satisfying all the key features desired by the device designer.
However, the advantages go well beyond the improvements of device
isolation (see Table II). The widths of the isolation grooves are
not dictated (as previously) by process limitations, but rather by
the state-of-the-art lithography available at the time. Thus,
extendibility with future refinements in lithography is assured.
Depth limitations for the isolation regions also do not exist.
Furthermore, in its ultimate application, the trench isolation
process eliminates the requirement for both the subcollector and P+
diffusion subisolation masks, as well as the P+ diffusions (see
Figure 2). Instead, a blanket subcollector is diffused into the
substrate across the whole surface of a wafer. This eliminates
another key bipolar device process concern – epitaxial
autodoping (24,25). The autodoping is sensitive to device pattern
densities and typically must be controlled or adjusted with each
new design. The use of a blanket subcollector requires the
development of only one epitaxial process parameter set. The blanket
subcollector concept also eliminates the need for a subcollector

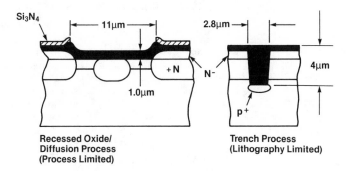

Figure 2   Details of oxide/diffusion and trench isolation
           structure. Line width equals 2.5 μm.

           Table II.   Trench Isolation Technology Features
                       (for Bipolar Devices)

-    Blanket Subcollector
-    Avoidance of Lateral Epitaxial Autodoping
-    Elimination of Sub-isolation Diffusion
-    Single Mask Isolation and Subcollector Definition
-    Absence of Bird's Beak
-    Planar Surfaces
-    Improved Device Density - Lithography Controlled
-    Improved Device Performance
-    Butting of all Device Elements
-    "Unlimited Isolation Depth"

---

reoxidation step, previously required for assuring proper positioning of the P+ subisolation mask. Hence, the overall epitaxial process becomes much simpler with the trench isolation method.

Another process simplification is the fact that the Si_3N_4 as an oxidation masking film is not used. The absence of this film permits the fabrication of quite planar surfaces, absent of bird's head and beak features. With the ability to butt all device elements against the trench oxide, one is, for example, able to realize improvements in bipolar device densities and performances of up to 30-40% in comparison to present isolation schemes. Further improvements are limited, mainly because of the state-of-the-art device interconnection technology.

The overall process sequence consists of four basic process steps (refer to Figure 3):

- Trench Formation (RIE)
- Trench Filling (CVD)
- Surface Planarization (Resist)
- Excess Material Removal (RIE)

These process steps, as practiced for the trench isolation process, are either new or modifications or extensions of previously practiced processes used for other aspects of device fabrications. Consequently, several of these steps did require general process development on an individual basis, prior to their integration into the overall process. Some of the efforts have resulted in new basic knowledge in the area of reactive ion etching (RIE) and chemical vapor deposition, surface planarization with resist materials, and thermal oxidation or nonplanar silicon surfaces. The author has previously presented various aspects of these process activities (4-8), as applied to the bipolar device technology.

Trench Formation

Optimum device isolation is not only restricted to the deep trench isolation between devices, but also calls for shallow trenches to allow isolation between the base and the subcollector reach-through diffusion. This trench reaches down to the subcollector only. A third trench shape option is that of wide, deep trenches. These would be formed outside of the immediate device areas and serve (after filling with SiO_2) as the bed on which to locate the wiring lines which interconnect the different devices and circuits. Locating wires on such SiO_2 areas reduces the wiring capacitances significantly.

Crystallographic silicon tends to be etched isotropically when using traditional wet chemistry etchants. This results in different groove shapes on different crystallographically oriented surfaces. In the case of (100) oriented surfaces, these grooves can be rounded or V-shaped (see Figure 4). The desired near vertical and deep grooves can however be achieved routinely with the more recently introduced reactive ion or plasma etching technology, provided the correct chemistry and etching conditions

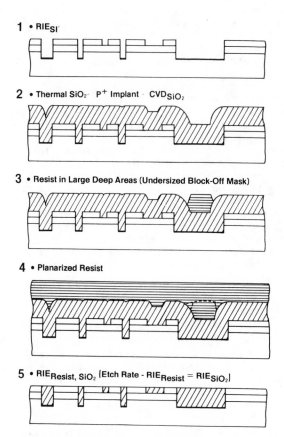

1 • RIE$_{Si}$

2 • Thermal SiO$_2$· P$^+$ Implant · CVD$_{SiO_2}$

3 • Resist in Large Deep Areas (Undersized Block-Off Mask)

4 • Planarized Resist

5 • RIE$_{Resist, SiO_2}$ [Etch Rate - RIE$_{Resist}$ = RIE$_{SiO_2}$]

Figure 3   Process sequence to achieve planar surfaces over narrow and wide trenches.

are established. Both fluorinated- and chlorinated-based
chemistries have been found to meet the etching needs. For the
present work, a gas mixture of Cl_2/Ar has been used to generate
very deep and vertical trenches in (100)-oriented silicon surfaces
(see Figure 5). In fact, this etching chemistry allows trench
shape tailoring from vertical trenches to heavily tapered ones
(26). Figure 6 is an example of a combination of both deep and
shallow trenches, whereas Figure 7 shows deep, wide trenches. The
formation of all these different types of trenches were formed
using similar RIE conditions.

The etchant mask for the Cl_2/Ar chemistry is SiO_2. The
lithographically defined window in the SiO_2 mask dictates the width
of the vertical trenches, whereas mask erosion tends to widen the
trenches and therefore defines the tapered trenches. The etch rate
ratio between Si and SiO_2 can be varied by reactor parameter
changes. With an etch rate ratio of about 20:1 ($Si:SiO_2$), an SiO_2
film thickness of about 200-300 nm will suffice to serve as a mask
for etching 4-5 μm deep trenches (see Figure 5). The etch rates
for Si are in the range of 50-100 nm/min.

Some of the etch characteristics of silicon are dependent on
the resistivity type, particularly highly doped regions, such as
arsenic-doped subcollectors. Using higher etch rates, these
diffused regions can exhibit significant lateral underetching (26).
Experiments have shown that this can be attributed to an enhanced
thermal heating effect in those regions, becoming accentuated
during high etch rate conditions. A compromise of the overall etch
rate of silicon is therefore necessary to maintain proper trench
shape control.

Trench Filling - Initial Processing

The trench may be filled with any of several different materials.
In the case of device applications, the filling material is some
form of dielectric material, such as oxides, polymers, or
polycrystalline silicon. (Metal filled trenches could serve as
buried interconnecting device lines). As suggested previously,
generally two types of trenches may have to be filled up - narrow
ones for device isolation, and wider ones to benefit other circuit
functions. The process requirements are somewhat different for
each. However, independent of the trench width, the initial
processing is the same.

Of first importance is the creation of a very high quality
surface oxide lining the trench walls to assure a good Si/SiO_2
interface with low surface state densities. The typical deposition
materials do not meet these criteria. Therefore, the initial
trench processing consists of a thermal oxidation cycle to form
about 100-200 nm of SiO_2. Any further thermal oxidation proves to
be of no real benefit; in fact, it is not even desirable,
particularly as a full trench-filling methodology (see Figure 8).
This is for the following reasons:

-   Thermal cycles become impractically long.
-   Method requires protective surface film to assure trench
    oxidation only.
-   Vertical trench shape is lost.

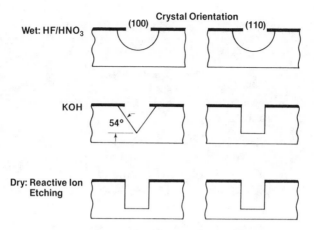

Figure 4   Comparison of silicon etchants for (100) and (110)
orientations.

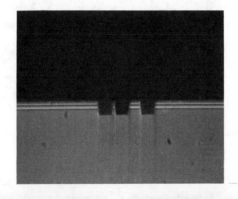

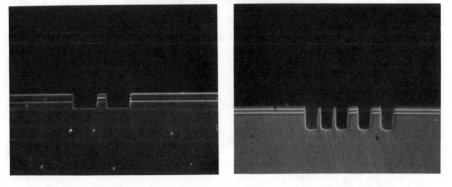

Figure 5   Silicon trenches (grooves) created with Cl_2/Ar
reactive ion etching. The lines indicate the buried
subcollector.

a   b

c

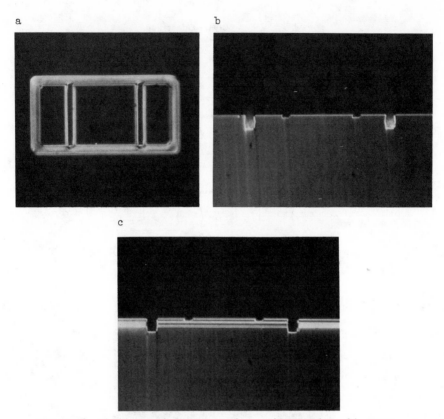

Figure 6   Shallow and deep trenches; a) top view, b) cross section, c) cross section with N+ subcollector decorated.

Figure 7 Wide trenches created with reactive ion etching.

- Isolation width becomes twice that of the starting trench width.
- Top center region of oxidized trench does not effectively close off (Figure 9).
- Final structure may exhibit defects (dislocations) in silicon surrounding the trench area.
- Wafer warpage.

Even the 200-nm thick thermal oxide over the sharply changing trench contours needs to be formed under proper processing conditions. As can be seen in Figure 10, the crystallographic defect generation tends to be initiated at the sharp corners of the trench geometry, and is sensitive to both the oxidation temperature and oxidation thickness. It is also quite dependent on the exact trench shape and any surface irregularities. Thus, only a limited thickness of thermal oxide should be grown at a particular temperature, with higher temperatures allowing thicker film growths (27). Similar observations have been made by Parekh (28) for less vertical grooves, which were masked by an oxide/nitride layer, by Fang, et al. (20) for SWAMI-type structures, and by Takano and Kozuka (29) and Chiang et al. (30) for silicon grooves. Because of the tendency for dislocation generation, Hayasaka et al. (31) have opted to form sloped geometries at the top of their trenches.

Prior to the final trench filling, a trench implantation (boron for P-type substrates) is performed. This is to ensure that the P-type substrate does not invert during subsequent processing and cause device-to-device leakages. The implant is directed vertically into the trench. Under these conditions the trench sidewall oxide acts as an efficient mask, while the implant in the bottom can reach into the substrate and form a P+ region. The remaining SiO_2 film, which served as the RIE mask, also acts as an implant mask at the wafer surface. The implant energy and dose are adjusted for the thermal oxide thickness.

Trench Filling - Completion with Chemical Vapor Deposition

The completion of the trench filling may be done by one of several techniques, and as stated previously, with one of several materials. However, the most obvious approach is the use of the chemical vapor deposition (CVD) technique, and the primary choices are SiO_2 (CVD_{SiO_2}) or polycrystalline silicon ($CVD_{poly-si}$).

CVD techniques have been found to be excellent for coating surfaces conformally, especially for most nominal surface configurations encountered in the present device structures. Only limited data exists for CVD coatings over steep and dominant surface steps or into narrow deep grooves (trenches). A recent report by Breckel and Bloem (32) on polycrystalline silicon depositions into trapozoidal and V-groove configurations indicates that the CVD growth characteristics can change dramatically with growth temperature and growth rates. The groove dimensions, used by these investigators, are still larger (i.e., wider) than those intended for the trench isolation requirements. Consequently, the data is not completely applicable for the present need. Smeltzer's

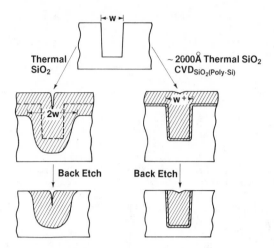

Figure 8   Comparison of a silicon trench fully oxidized or filled in with CVD.

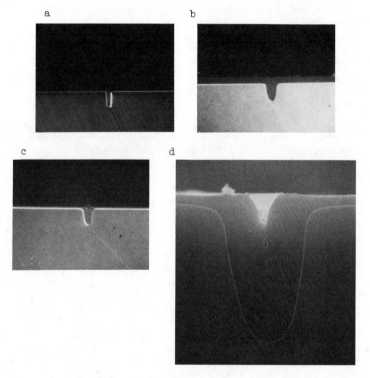

Figure 9   Examples of thermally oxidized trenches; a) etched trench, b) thermally oxidized, c) backetched, d) details of c).

a

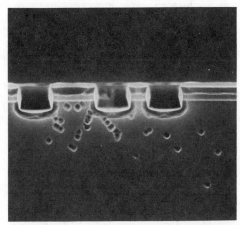

b

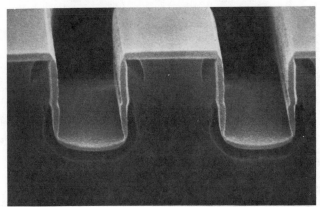

Figure 10 Oxidized trenches treated with a dislocation
          generating etchant; a) 900°C oxidation with
          dislocations, b) 1000°C oxidation without
          dislocations (gouging around trench is due to stress
          in silicon).

report, relating more specifically to trench filling, discusses
silicon epitaxial depositions into silicon trenches (33). With
adjustments in the chemical equilibrium via changes in the Si/Cl
vapor phase ratio, he was able to achieve zero growth rates on a
horizontal wafer surface, while maintaining acceptable deposition
rates in deep and narrow trenched regions. He thereby filled up
the trenches preferentially. As the trenches became filled, the
growth rate equilibrium shifted gradually towards a no-growth
condition. This allowed simultaneous filling-up of different
trench depths, while the wafer surface experienced no growth at
all. These conditions could possibly be translated for
polycrystalline silicon trench filling; however, similar growth
conditions would be difficult to achieve for SiO_2 film depositions,
unless significant amounts of Cl_2 incorporation into the SiO_2 film
could be tolerated. This is not the case. Based on the current
state of CVD processing, therefore, additional process development
had to be pursued to realize successful trench filling.
Furthermore, two types of trenches may be required to be filled –
narrow ones and wide ones.

Narrow Trench Filling – Completion with CVD SiO_2

Successful CVD SiO_2 trench filling requires a focus on the
understanding of the deposition characteristics of a binary vapor
phase reaction system. Typically, the oxygen and silicon
containing gas species exist in a high oxygen/silicon vapor phase
ratio for this type of deposition system. This implies that in
order to achieve a stoichiometric SiO_2 deposition, the
silicon-containing gas species tends to deplete much faster than
the oxygen-containing species, the reaction being controlled by a
gas diffusion mechanism. Hence the vapor phase ratio near the
growth surface quickly increases in comparison to its original
input level. In the case of a planar surface the resupply is such
that a stable growth condition can be established. A cursory
evaluation of the influence of HCl on the deposition
characteristics was made at 900°C with a $SiH_4/CO_2/H_2$-based
chemistry. Over a range of 1-3% HCl in the vapor phase, no
influence was observed – either negatively or positively. This
direction was not further pursued, since higher HCl concentrations
would result in too much Cl_2 incorporation into the SiO_2 film.
    The initial CVD SiO_2 depositions into deep, narrow trenches and
over sharp steps were made at 780°C with a $SiH_4/N_2O/N_2$ based
chemistry. The data identified several important aspects of
CVD SiO_2 contouring characteristics and trench filling. Figure 11
indicates the excellent contouring characteristics over sharp and
narrow ridges, the ridges being as narrow as 100 to 500 nm. A
detailed analysis indicates a higher growth rate at the top than on
the side of the ridge wall. The same effect occurs in trench
geometries, as shown in Figure 12. The final angle of the oxide
wall tends to become more negative as the deposition progresses
(Figure 12c). The change of the oxide angle varies also with the
starting trench angle; the specific relationship is indicated in
Figure 13.

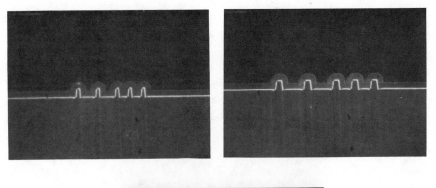

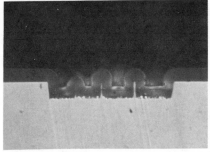

Figure 11 CVD SiO_2 film depositions over sharp surface steps.

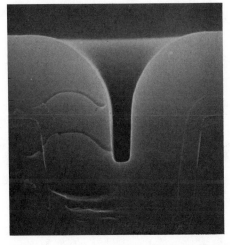

Figure 12 Details of a trench after initial thermal SiO_2 and CVD SiO_2. Note angle of trench wall and inside CVD SiO_2.

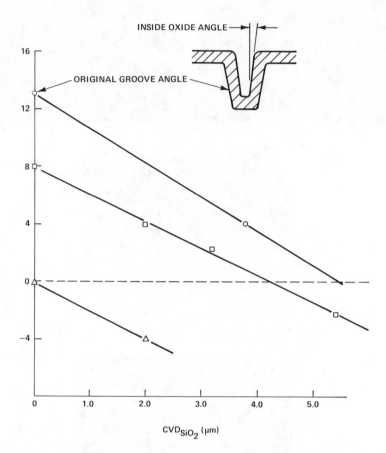

Figure 13 Relationship between inside CVD oxide wall angle and $^{CVD}Si0_2$ thickness. $^{CVD}Si0_2$ at 780°C.

The above observations apply to trenches with relatively wide dimensions. When the aspect ratio between the trench width to depth becomes smaller, the deposition can possibly result in physical overgrowth. This will leave a narrow void in the center of the filled trench, as is shown in several examples in Figure 14. The effect occurs more readily for trenches with vertical walls (Figure 13), and with certainty for trenches with convexed sidewalls.

Since the deposition mechanism is diffusion controlled, the results of the 780°C depositions led to the evaluation of other growth temperatures and other based chemistries, specifically as follows:

Chemistry	Temperature	Reactor System
SiH_4/O_2	400°C	LPCVD
$SiH_4/N_2O/O_2$	780°C	Atm (Horizontal)
$SiCl_2H_2/N_2O$	900°C	LPCVD
$SiH_4/CO_2/H_2$	900°C	Atm (Barrel)
$SiH_4/CO_2/H_2$	1000°C	Atm (Barrel)

The main focus of this study was to assess the relationship between the horizontal and vertical deposition rates for these different deposition systems. Figure 15 is a plot of the horizontal/vertical deposition ratio as a function of the deposition temperature. At about 900°C this ratio becomes 1.0, whereas at 400°C the value is as high as 1.6. Variation and optimization of the growth conditions at each of the temperatures would tend to change the reported aspect ratio to some degree. However, the data clearly indicate the contouring efficiency of CVD_{SiO_2} to be independent of a particular chemical system as well as of a specific reactor system (i.e., atmospheric, low pressure) - at least it is not a first-order effect to these parameters. Consequently, low temperature depositions are not acceptable for efficient trench filling applications.

These data remain consistent with the gas diffusion reaction model. Thus, in the area of a deep and narrow trench, an adequate resupply of the silicon gas species to the deposition surfaces within the trench is not met as readily as on the top of the wafer's surface. Inevitably, the silicon gas specie will tend - because of its proximity - to deposit on the upper portion of the trench sidewalls. Hence, the growth rate near the top of the trench is faster than further down towards the bottom, a condition prone for eventual physical overgrowth (or closure) and void formation. A significant lowering of the vapor phase O_2/Si ratio tends to cause general deposition problems in terms of growth rates (very low) and poor uniformities for typical batch sizes. Consequently, this approach is not fruitful.

The gas phase diffusion of the Si species becomes sufficiently low at 400°C so that little lateral growth occurs in the grooves.

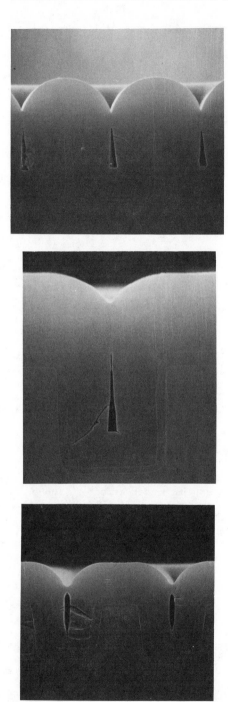

Figure 14 Examples of CVD SiO_2 physical growth into trenches at
low CVD temperatures.

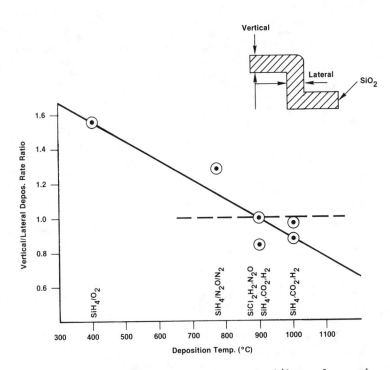

Figure 15 Relationship between the vertical/lateral growth rate ratio and the CVD$_{SiO_2}$ growth temperature.

The gas diffusivities improve with higher deposition temperatures, and at about 900°C, the lateral and vertical growth rates tend to become nearly equivalent and good trench filling occurs (see Figure 16). However, in the very last phase (i.e., just prior to final trench closure) the gas stoichiometry appears to remain inadequate for good quality oxide growth. This aspect is shown in Figure 17, where the cross section of a filled trench is treated with an oxide-sensitive etch. The center region etches at a higher rate than the other oxide, thereby creating a void. Since this "poor" oxide in the center region is a direct result of an inappropriate gas phase stoichiometry, attempts have been made to improve it by means of heat annealing. The results were more successful at higher temperatures, particularly for 1000°C anneals (8). Alternatively, a more tapered trench allows more efficient trench filling and closure. Incidentally, it should be noted (and as seen in the microphotographs) that the CVD_{SiO_2} deposits do not only occur in the trench area, but also all across the wafer's surface.

Narrow Trench Filling - Completion with $CVD_{Poly-Si}$

Experimentally, trench filling with $CVD_{Poly-Si}$ is simpler in comparison to filling with CVD_{SiO_2}. In fact, the processing is relatively routine. Since the deposition is based on a single gas phase supply system, normal low pressure deposition conditions at 650°C are quite adequate for this application. If greater control over the crystallinity of the polycrystalline silicon is desired, more attention must be given to establishing the appropriate growth temperatures. In either case, the trench filling itself is readily achieved, provided that the trench is near vertical or tapered slightly outward (Figure 18). As was found in the case of CVD_{SiO_2} (although more pronounced) the filling of slightly concaved trenches resulted in some amount of void formation, as seen in Figure 18b. Also, the required deposition thickness is dictated by the trench width; that is, slightly more than one-half of the trench width. Trench filling with polycrystalline silicon has been reported in the literature (30,31,34-37), although process details have generally not been documented.

Wide Trench Filling - CVD_{SiO_2}

The process requirements for filling up large area trenches (or more properly, wide depressions) are somewhat different than those for narrow trench filling. From a process commonality point of view the dielectric would tend to be SiO_2, not polycrystalline silicon. As stated previously, the CVD_{SiO_2} does contour quite well over major surface steps. However, since the depression width is much larger than the depression depth, a "closing-up" of the depression similar to that of a narrow trench does not occur. Instead, the depression must be literally filled up with a thick enough deposition. The overall step height of the depression is retained, since the deposition also occurs on the top of the wafer. Thus the

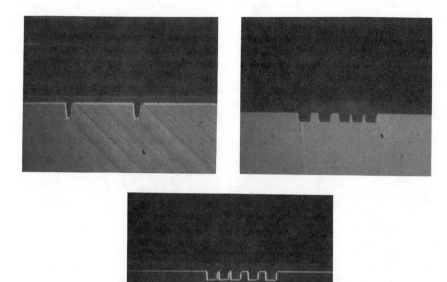

Figure 16 CVD$_{SiO_2}$ filled trenches at 900°C.

a   b

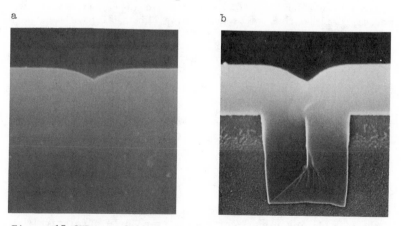

Figure 17 CVD$_{SiO_2}$ filled trenches; a) cross section untreated, b) cross section decorated to delineate SiO_2 (center seam is due to "poorer" quality SiO_2).

surface step is displaced upward by the thickness of the deposition. To assure wide area trench filling, the CVD_{SiO_2} thickness is planned to be slightly more than the trench depth (normally about 1.0 μm, but the depth could be more, if desired), so that the top surface of the SiO_2 film is above the surface of the adjacent silicon substrate (Figure 19). These thick deposits will well overcoat any narrow trenches, so well in fact, that they would exhibit minimal surface contouring. The ground rule that the trench shape is to be slightly tapered also applies to wide trenches.

Post-CVD Surface Planarization

The planarity of the surface above a narrow CVD-filled trench may be quite good, provided a sufficient amount of CVD material has been deposited. The specific planarity will depend directly on the trench width and the CVD film thickness. This relationship is shown in Figure 20. Surface planarity will not be achieved for wider trenches. Such major surface irregularities cannot be tolerated as a final surface structure, and a post-CVD surface planarization technique must be employed to assure a sufficiently planar surface prior to a subsequent back-etching process step.
     Two techniques have been discussed in the literature. Higuchi et al. (38) utilized a leveling method in which a resin film is deposited on the wafer and flattened with a copper plate prior to the resin solidification. The obvious concern with this technique is whether sufficient planarity can be achieved with the typical product wafer, and whether the resultant process tolerances are acceptable for advanced device design requirements. An alternative method has been reported by Adams and Capio (34), who applied a photoresist film onto a wafer containing abrupt steps. The resist layer tends to smooth the surface with the help of a bake cycle. A subsequent back-etching with a RIE technique results in a replication of that smoothened resist surface in the underlying material which originally exhibited the abrupt surface steps. The back-etching requires equivalent or nearly equivalent etch rates between the resist and underlying material. This method does not assure completely planarized surfaces, but rather gently changing surface contours. In concept, the technique can also apply to planarizing depressions. The specific process requirements will again be somewhat dictated by whether one is working with narrow or wide trenches. These results have been discussed previously (7).

Planarizing Narrow Trenches

As is evident from Figure 20, to achieve an adequate planar surface with the CVD filling - even for narrow trenches - a relatively thick film deposition is required. The presence of this thick film is not very attractive, especially in the light that such films must be removed from the top of the wafer surface. With the aid of the planarizing organic material (resist or other polymer), the need for thick CVD depositions can be reduced significantly. Under such circumstances, the deposited film thickness needs only be enough to adequately fill the trenches; that is, the bottom of the

a   b

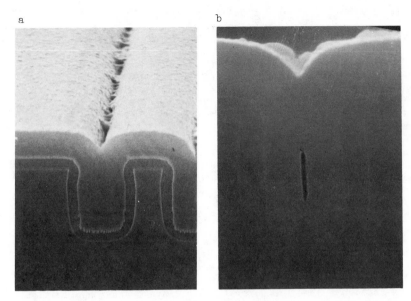

Figure 18 CVD$_{Poly-Si}$ filled trenches after thermal SiO$_2$ liner;
a) fully filled, b) physical overgrowth.

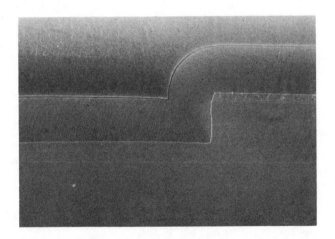

Figure 19 Wide trench filled up with CVD$_{SiO_2}$.

crevice in the center of the filled trench should be above the wafer's surface. This remaining surface contour can be covered and made planar with a single coat of planarizing resist, as seen in Figure 21. Similar good planarity has also been observed for sets of trenches spaced very close together. However, some degree of application optimization is required for specific combinations of trench densities, pattern varieties and trench widths. Resist-filled trenches (those wider than about 3.5 μm) tended to exhibit some slight surface troughing in the center region of the trench. This feature became more and more prevalent with further widenings of the trench width. At a width of about 10 μm (after trench filling), satisfactory resist planarity is lost. The resist thickness required for surface planarizing these nominal dips and troughs of narrow trenches tends to be about 1.0 μm. However, the optimum thickness will of course depend on the resist type, application conditions, bake cycles, and to some degree, also, on trench pattern densities.

Planarizing Wide Trenches

In order to achieve satisfactory planar surfaces over wide trenches (>10 μm) on a wafer, a double resist coating method was employed. (Refer to Figure 3). The initial resist layer, after application, is exposed with an undersized block-off mask, and, with appropriate processing, will leave the resist only in the depressions of the wide area trenches. The resist thickness of this layer should be about equivalent to, or slightly more than, the depth of the depression - in the present case about 4-5 μm. The resist thickness is also dictated in part by the dimensional difference between the block-off mask size and the trench depression area. The wafer is next given a bake cycle in which the resist flows within the depressions and more evenly "fills" those depressions (Figure 22). Only under optimum conditions would this result in very planar surfaces (Figure 22d). More typically, some slight under or overfilling would occur with some degree of surface undulations. These become corrected by a second resist coating, quite similar to the single coating of narrow trenches.

  Process optimization of the two-coat process should place focus on achieving as good a planarity as possible with the first block-off resist. If this first resist layer does result in nominally flat surfaces, the final planarity for 100-μm wide trenches has been found to be as good as 300 nm (Figure 23). The second resist layer thickness tends to be of the order of 1.0 μm; however, the specifics again depend on similar parameters as discussed previously. The second resist coating for the wider trenches would serve as the single coat resist for the narrower trenches, provided both types are present on the design layout.

Excess Material Removal

The combination of the CVD trench filling and the surface planarizing material (PM) results in a total material build-up on top of the wafer's surface that could be several μm thick. The specific thickness will of course depend on the type of trenches that happen to be present on the wafer. If a wafer contains only

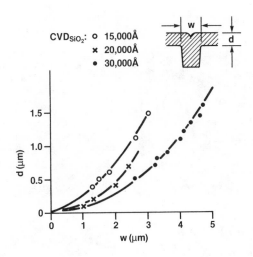

Figure 20 Relationship between surface dip and trench width for
different CVD_{SiO_2} thicknesses.

Figure 21 Single coat of resist over relatively narrow (< 10 um)
trenches.

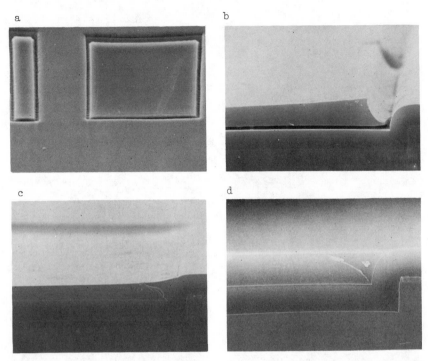

Figure 22 Details of block-off resist in wide area trenches; a) top view, b) cross section (delamination caused by sample treatment), c) after bake cycle, d) same as c, but more fully filled.

narrow trenches, the excess layer may only be as much as 3 μm (2 μm SiO_2 or polycrystalline silicon and 1 μm of planarizing material). However, if large area trenches with depths of 4-5 μm are present, then the total thickness above the wafer's surface could be as much as 6-7 μm (4-5 μm of SiO_2 and 1-2 μm of planarizing material over the non-trenched area or any narrow trenches). In the large area trenches, the resist thickness would be 6-7 μm. Examples are seen in Figure 23.

This excess material build-up must be removed to re-expose the wafer surface, so that further device processing can proceed. The material removal is done by a back-etching technique using a RIE or plasma etching method. This approach is used in preference to that of traditional wet chemistry etching, primarily because of the greater degrees of freedom of selecting appropriate etching conditions. For this particular need, it is important to exercise etching conditions in which the etch rates of both the planarizing material (RIE_{PM}) and the CVD material (RIE_{CVD}) are very nearly equal. Such conditions would then allow a uniform removal of the overcoatings from everywhere across the wafer, whether it be just the PM or a combination of PM and CVD. If the CVD deposition and the PM coating uniformities are well controlled, the back etching will result in uniform material removal down to the silicon surface and leaving the filled-in trenches with a planar surface. A laser end-point detect scheme is used to control the etch cycle. The filled trenches will be nominally flush with the silicon surface. Figures 24 and 25 are examples of narrow and wide trenches processed through this sequence. In Figure 24a and b the etch rate of the CVD_{SiO_2} film was in this case slightly higher than that of the resist, and this results in the small surface step. Opposite etch rate conditions (resist etch rate higher) were used for the sample shown in Figure 24c. Notice the difference in the final planarity of these samples if they had been etched to completion. This suggests that if there is a non-equivalency in etch rates, it is preferable that it be in favor of the CVD film etch rate. Figures 24d, 25, and 26 give examples of back-etched samples of different groove widths and depths in which the etch rate conditions for SiO_2 and resist were very nearly equivalent.

Final Processing

The planarized structures are at this point ready for follow-on device fabrication processing. Due to the tendency of "poor" quality oxide in the center region of the filled narrow trenches, a heat annealing of the structure should be made. This improves the oxide quality so that it is nearly equivalent to the regular oxide. Alternatively, a CVD_{SiO_2} deposition over the back-etched surface can serve the same purpose, while at the same time it becomes the oxide masking layer for follow-on processing. Figure 27 shows a process sequence of etched trenches, filled trenches, and back-etched trenches.

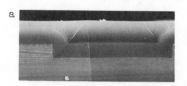

Figure 23 Examples of planar surfaces with the double resist coating method for very wide trenches; a) 36 μm wide, b) 110 μm wide.

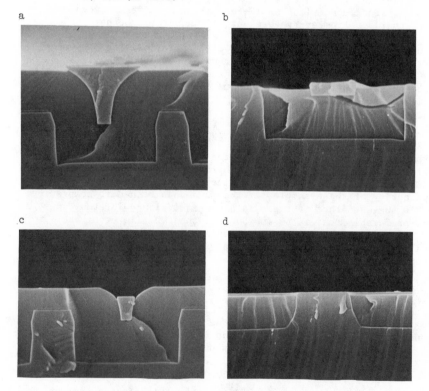

Figure 24 Details of RIE back etching of planarized resist and SiO_2; a) partially backetched with ER_{SiO_2} slightly higher than ER_{resist}, b) same as a) but fully backetched; c) partially back etched with ER_{resist} slightly higher than ER_{SiO_2}, d) fully back etched with equal etch rates.

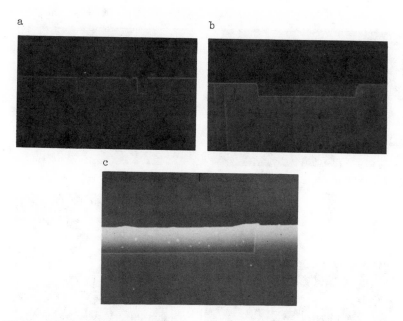

Figure 25 Examples of CVD SiO_2 filled trenches back etched with
RIE; a) shallow and deep, b)wide, c) details of b).

Figure 26 CVD SiO_2 filled and back etched wide trenches; a)
85 um wide, b) 110 μm wide.

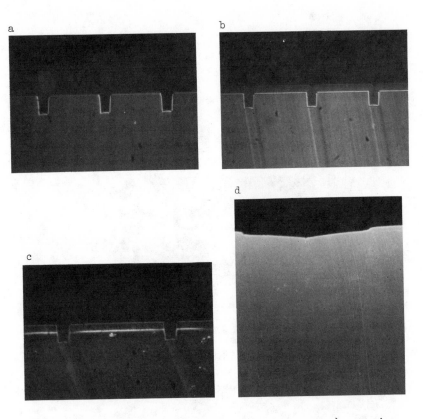

Figure 27 Sequence of device isolation trenches; a) reactive ion etched trenches, b) CVD-filled trenches, c) back etched with N+ subcollector decorated, d) details of c).

Conclusion

A new device-isolation concept has been presented along with
various aspects of the associated process components. This
isolation scheme has a number of advantages over other currently
practiced isolation methods, and it meets all the basic
requirements for achieving the high density, high performance
devices and circuits expected in VLSI designs. In fact, it makes
their fabrication much more achievable. Excellent electrical data
have been generated for devices and circuits incorporating the
trench isolation structure, and this form of device isolation
appears to have found quick acceptance by a host of other
investigators. This is evident by the dramatic increase over the
past few years in the number of published papers and presentations.
    All the early work appears to focus only on narrow trenches of
a single depth, mainly because of process simplicity. However,
Kurosawa, et al. (40) discuss oxide filling of wide shallow
trenches. The use and application of the trench technology is not
only restricted to bipolar device designs, but is also being
incorporated in FET device structures. As has occurred in previous
process technology developments, some degree of process variations
(in comparison to those presented here) are being pursued by
different investigators. For example, Goto et al. (34) report on
the use of a chemical mechanical back-polishing technique (instead
of the RIE back etching), and achieve excellent results.
    As the VLSI device fabrication efforts intensify over the next
few years, it is becoming very clear that the trench technology
will become a dominant feature and the trench isolation an integral
part in most of those VLSI device and circuit structures.

Acknowledgments

I would like to express my specific thanks to the following people
who, through their respective experimental assistance, actively
supported some of the development activities of the various process
elements: W. Ames, J. Anderson, H. Bhatia, J. Bondur, R. Broadie,
P. J. Burkhardt, C. Crimi, D. Dyer, C. D. Garrison, R. Gray,
T. Hansen, E. Hearn, C. T. Horng, G. Kaplita, B. Kemlage,
J. Lechaton, H. Lillja, A. Michel, R. Schwenker, R. Wilbarg and
B. Zingerman.

Literature Cited

1. I. Ishida et al. IEEE IEDM, Washington, 1979 p. 336.

2. D. B. Tang et al., IEEE Solid St. Circuits, Vol. SC-17, 925,
    October 1982.

3. S. Brenner et al., IEEE ISSCC83, 1983, p. 152.

4. H. B. Pogge, J. A. Bondur and P. J. Burkhardt, Recent
    Newspaper, The Electrochem. Soc. Meeting, Pittsburg,
    October 1978.

5. H. B. Pogge, Recent Newspaper, The Electrochem. Soc. Meeting, Boston, May 1979.

6. H. B. Pogge, ESSDERC 1980 Symposium, York, England, 1980.

7. H. B. Pogge, G. A. Kaplita and R. R. Wilbarg, The Electrochem. Soc. Meeting, Hollywood, Fl., October 1980, Abstract 309.

8. H. B. Pogge, AICE Meeting, Detroit, August 1981.

9. See, for example, papers in IEEE IEDM Meetings 1982 and 1983.

10. K. Lehovec, U.S. Patent #3,029,366, April 10, (1962).

11. E. H. Porter, U.S. Patent #3,260,902, July 12, (1966).

12. J. A. Appels et al., Philips Res. Rep., 25, 118 (1970).

13. E. Kooi et al., Philips Res. Rep., 26, 166 (1971)

14. D. Peltzer and H. Herndin, Electronics, 44, 52 (1971)

15. F. Morandi, IEEE IEDM, Washington, 1969, p. 126.

16. K. Y. Chiu et al., IEEE IEDM, Washington, 1982, p. 224.

17. J. C. Hui et al., IEEE Trans. Electron Devices, Vol. ED29, 554 (1982).

18. E. Kooi and J. A. Appels, Semiconductor Silicon 1973, H. R. Huff and R. R. Burgess, eds., The Electrochem Soc. Princeton, N.J., 860 (1973).

19. E. Bassous et al., J. Electrochem. Soc., 123, 1729 (1976).

20. R.C.Y. Fang et al., J. Electrochem. Soc., 130, 190 (1983).

21. U. S. Davidsohn and F. Lee, Proc. IEEE, 57, 1532 (1969).

22. K. E. Bean et al., Semiconductor Silicon 1973, H. R. Huff and R. R. Burgess, eds., The Electrochem. Soc., Princeton, NJ, 880 (1973).

23. H. Mukai et al., IEEE ISSCC77, 1977, p. 77.

24. H. B. Pogge, Vapor Phase Epitaxy, in Handbook on Semiconductors, Vol. III, Materials, Properties and Preparation S. Keller, ed., North Holland, N.Y., 335 (1980).

26. H. B. Pogge, J. A. Bondur and P. J. Burkhardt, J. Electrochem. Soc., 130, 1592 (1983).

27. P. J. Burkhardt, private communication.

28. P. C. Parekh, J. Electrochem. Soc., 125, 1703 (1978).

29. Y. Takano and H. Kozuka, Digest of Tech. Papers, 14th Conf. on Solid St. Devices, Tokyo, C6-3, 231 (1982).

30. S. Y. Chiang et al., The Electrochem. Soc. Meeting, Detroit, 1982, Abstract 174.

31. A. Hayasaka et al., IEEE IEDM, Washington, 62 (1982).

32. C. H. J. van den Brekel and J. Bloem, Philips Res. Rep., 32, 134 (1977).

33. R. K. Smeltzer, J. Electrochem. Soc., 122, 1666 (1975).

34. H. Goto et al., IEEE IEDM, Washington, 58 (1982).

35. A. Hayasaka and Y. Tamaki, JEE 19, 36 (1982).

36. R. D. Rung et al., IEEE IEDM, Washington, 237 (1982).

37. Y. Tamaki et al., Proc. 13th Conf. Solid St. Devices, Tokyo, 37 (1981).

38. H. Higuchi et al., The Electrochem. Soc. Meeting, St. Louis, 1980, Abstract 173.

39. A. C. Adams and C. D. Capio, J. Electrochem Soc., 128, 423 (1981).

40. K. Kurosawa et al., IEDM Digest, 384 (1981).

RECEIVED June 6, 1985

Solid-Liquid Equilibrium in Ternary Group III-V Semiconductor Materials

T. L. Aselage[1], K. M. Chang, and T. J. Anderson

College of Engineering and Chemical Engineering, University of Florida, Gainesville, FL 32611

A general approach to calculating solid-liquid phase diagrams of ternary Group III-V semiconductor material systems is presented in which the standard state chemical potential terms are evaluated separately from the liquid and solid solution activity coefficient terms. Three methods of determining a reduced standard state chemical potential change are given and their implementation is illustrated for the Ga-Sb system. Knowledge of the deviation from ideal behavior of the liquid mixture properties relative to the solid solution is important for calculating ternary phase diagrams and the calculation is demonstrated for the Al-Ga-Sb system.

The compounds formed by the Group IIIA elements of the periodic table, Al, Ga and In, with the group VA elements, P, As and Sb, have the potential to be extremely important semiconductor materials. The attractiveness of Group III-V compounds as electronic materials lies in the variability of electrical properties among the different compounds and the fact that these properties are often superior to those found in Si.

For devices which require high operating speeds or frequencies, several III-V materials offer a significantly larger electron mobility than Si. Indeed, for the high mobility compounds InAs and InSb the improvement is greater than an order of magnitude. For optical device applications, it is advantageous to utilize a material with a direct bandgap; the binary compounds GaAs, InP, GaSb, InAs and InSb have direct bandgaps, while Si has an indirect bandgap. Because of the more complex native point defect structure of III-V materials compared to Si, most III-V compounds can be made semi-insulating by controlled processing.

[1]Current address: Sandia National Laboratories, Albuquerque, NM 87185

In integrated circuits this results in lower power consumption and reduced parasitic capacitances.

Perhaps the most important property of III-V materials, however, is the ability to form completely miscible solid solutions for most of the multicomponent systems. This is accomplished either by the substitution of an atom on the Group III sublattice, e.g. Ga, by another Group III element, e.g. In, or by the substitution of an atom on the Group V sublattice, e.g. As, by another Group V element, e.g. P. By the variation of the solid solution composition, the electrical properties of the solid are affected. This affords the device designer the ability to tune the electrical properties of the solid to fit the particular device requirements. In most applications, the designer will specify the bandgap energy to give the desired electrical characteristics and the lattice parameter to permit the growth of device quality epitaxial layers. To independently vary both of these material parameters often requires a quaternary system. Primarily because of simpler processing, certain Group III-V ternary solid compositions have received attention. In some systems the compositional degree of freedom in the ternary solid solution is used to match the lattice parameter to that of the substrate (e.g. $Ga_{0.47}In_{0.53}As$ lattice matched to InP). In $Ga_xAl_{1-x}V$ systems, however, the binary compounds have nearly identical lattice parameters. For these systems, the compositional degree of freedom can then be used to vary the bandgap energy, and therefore the electrical properties.

Ternary Solid-Liquid Equilibrium

Many of the processing steps in the fabrication of devices containing III-V materials involve the interfacial contact of a liquid phase with a solid phase. Examples of such processes are the melt or solution growth of bulk single crystals, the epitaxial growth of thin films from solution, and melt purification processes such as zone refining. As an equilibrium boundary condition is often implied at the solid-liquid interface, knowledge of the phase diagram for the material system is essential in the analysis of these processes. In addition, the thermodynamic behavior of III-V solid solutions is important in the analysis of processes that involve a solid-vapor interface such as chemical vapor deposition. A description of the solid solution behavior for III-V systems is often obtained from a reduction of available solid-liquid equilibrium data.

A general III-V solid-liquid ternary phase diagram is depicted in Figure 1. The two III-V binary systems, A-C and B-C, show similar behavior. Each system forms an equimolar compound, AC or BC, which melts at a higher temperature than either of the pure elements (except for the InSb-Sb case). The binary phase diagram consists of two simple eutectic systems on either side of the compound (e.g., the A-AC and the AC-C systems). The third binary phase diagram represents solid-liquid equilibrium between elements from the same group. In Figure 1 the A-B portion of the ternary phase diagram is depicted as being isomorphous

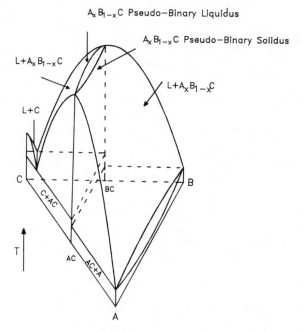

Figure 1.   Group III-V solid-liquid ternary phase diagram.

though other structures exist. The plane of equal number of atoms from the Group III and V columns contains the line compounds AC and BC and represents the $A_xB_{1-x}C$ pseudobinary phase diagram. The solid-liquid equilibrium region of the phase diagram is isomorphous for all but one of the possible 18 III-V ternary systems. In systems with a large difference in atomic sizes, miscibility gaps in the solid solution occur (not depicted in Figure 1). The other prominent feature of the ternary phase diagram shown in Figure 1 is the eutectic valley formed at compositions rich in C. The two-phase field, $A_xB_{1-x}C$ plus liquid, is the most important segment of the phase diagram for processing considerations.

The solid-liquid equilibrium state of the A-B-C ternary system is calculated by equating the temperature and pressure of each phase as well as the chemical potentials of each of the species present in both phases. In addition to these equations, a constraint of stoichiometry is placed on the solid solution; the sum of the mole fractions of the Group III elements must be equal to the sum of the mole fractions of the Group V elements. Because of this constraint, the chemical potentials of the three species are not independently variable in the solid. The ternary solid solution $A_xB_{1-x}C$ can be treated as if it were a binary solution of components AC and BC. The requirement of equal chemical potentials of each of the species present in both phases then becomes

$$\mu^s_{IC} = \mu^l_I + \mu^l_C \qquad I = A,B \qquad (1)$$

where I and C refer to the liquid phase components, and IC refers to the solid phase component.

By imposing the constraint of stoichiometry on the solid, the effect of the point defect structure is neglected. The binary III-V compounds are not strictly line compounds; some nonstoichiometry is present. The calculations of Hurle (1) and Edelin and Mathiot (2) indicate that the region of nonstoichiometry is quite small (less than 10^{-3} mole fraction) in GaAs and GaSb. Brebrick (3) has discussed the variation of the Gibbs energy of the solid for such nearly stoichiometric compounds, demonstrating that this variation is negligible for compounds with homogeneity gaps of less than 1% in mole fraction. Thus, although the chemical potential of a component in the solid varies considerably between the limits of stoichiometry, the sum of the element chemical potentials in the binary solid is nearly constant. This constant value for the sum of the solid element chemical potentials is the term μ^s_{IC} in Equation 1.

The chemical potential of component i in phase p, μ^p_i, can be related to a standard chemical potential, $\mu^{o,p}_i$, in terms of the activity coefficient, γ^p_i, and mole fraction, x^p_i, according to

$$\mu^p_i = \mu^{o,p}_i + RT \ln \gamma^p_i x^p_i \qquad (2)$$

The insertion of Equation 2 for each term in Equation 1, followed by algebraic manipulation, results in the following implicit

expressions for the liquid and solid equilibrium mole fractions of
species A and AC

$$
x_A = \frac{\frac{1}{x_C} - (1-x_C)\ \Gamma_{BC}\ \exp(-\theta_{BC})}{\Gamma_{AC}\ \exp(-\theta_{AC}) - \Gamma_{BC}\ \exp(-\theta_{BC})} \tag{3}
$$

and

$$
x_{AC} = \frac{x_C(1-x_C) - \frac{1}{\Gamma_{BC}}\ \exp(\theta_{BC})}{\frac{1}{\Gamma_{AC}}\ \exp(\theta_{AC}) - \frac{1}{\Gamma_{BC}}\ \exp(\theta_{BC})} \tag{4}
$$

Two types of variables appear in Equations 3 and 4, θ_{IC} and Γ_{IC}.
The variable θ_{IC} is a reduced standard state chemical potential
difference and is defined by

$$
\theta_{IC} = \frac{\mu_{IC}^{o,s} - \mu_I^{o,l} - \mu_C^{o,l}}{RT} \qquad I = A,B \tag{5}
$$

The term Γ_{IC} is the ratio of liquid phase to solid phase activity
coefficients and is given by

$$
\Gamma_{IC} = \frac{\gamma_I \gamma_C}{\gamma_{IC}} \qquad I = A,B \tag{6}
$$

The problem of describing the ternary III-V phase diagram is
reduced to selecting the standard states, determining the temperature and pressure (negligible) dependence of θ_{IC} and determining
the temperature, pressure (negligible) and composition dependence
of Γ_{IC}.

Reduced Standard State Chemical Potential Change

With the standard state for each component chosen as the pure component in the phase of interest and at the temperature of
interest, Chang et al. (4) have discussed three thermodynamic
sequences for the calculation of the reduced standard state chemical potentials. The pathways for each sequence are shown in Figure 2.

In method I (Figure 2a) the temperature of the solid compound
IC is raised from the liquidus temperature, T, to its melting temperature, T_m. The solid is then melted, forming a stoichiometric
liquid, which is cooled to the liquidus temperature. Finally, the
stoichiometric liquid is separated into the pure liquid elements.
The dashed line in Figure 2a indicates that these last two steps
involve a stoichiometric liquid below its normal freezing point,
i.e. a subcooled liquid. The sum of the Gibbs energy changes for

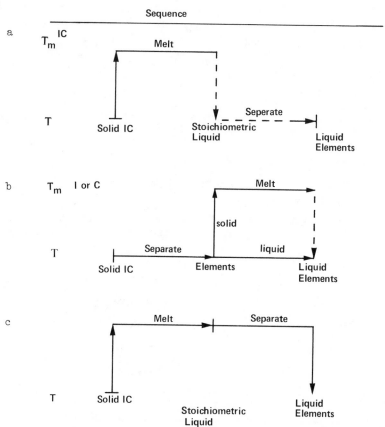

Figure 2.   Three thermodynamic sequences for evaluating the reduced standard state chemical potential change.

each of the steps in the process provides the standard state chemical potential difference, which is given by

$$\theta_{IC} = \ln\left[a_I^{sl}(T)a_C^{sl}(T)\right] - \frac{\Delta S_m(IC)}{R}\left(\frac{T_m}{T} - 1\right) + \frac{1}{RT}\int_T^{T_m}\int_T^{T_m}\frac{\Delta C_p(IC)}{T}dT^2$$

(7)

Here $a_i^{sl}(T)$ is the activity of component i in a stoichiometric liquid at the liquidus temperature, T, $\Delta S_m(IC)$ is the entropy of fusion of compound IC at the melting temperature, T_m, and $\Delta C_p(IC)$ is the difference in heat capacity between the stoichiometric liquid and the solid compound. This sequence is the same as that proposed for binary III-V systems by Vieland (5).

In method II, (Figure 2b), the solid compound IC is separated into the pure elements at the liquidus temperature, T. The Gibbs energy change for this step is simply the Gibbs energy of formation of the compound at the liquidus temperature. If the pure elements are liquids at this temperature, it is also proportional to the standard state chemical potential change. If one or both of the elements is a solid, a process of raising the temperature to the melting one, melting, and cooling to the liquidus temperature must be carried out. In this case, however, the sequence is performed on the pure element rather than the stoichiometric mixture. The dashed line in Figure 2b indicates that the final step in the process involves subcooling one or both of the pure liquids from their melting temperatures to the liquidus temperature. With this sequence the expression for the standard state chemical potential difference is

$$\theta_{IC} = \frac{\Delta G_f^o(T)}{RT} - \sum_{i=1}^2\frac{K_i}{RT}\left[\Delta S_m(i)(T_m^i - T) - \int_T^{T_m^i}\int_T^{T_m^i}\frac{\Delta C_p(i)}{T}dT^2\right]$$

(8)

where $\Delta G_f^o(T)$ is the Gibbs energy of formation of the compound at the liquidus temperature, T, $\Delta S_m(i)$ is the entropy of fusion of pure component i, T_m^i is the melting temperature of pure component i, $\Delta C_p^{(i)}$ is the difference in heat capacity between pure liquid i and solid i, and K_i is zero if i is a liquid at the temperature T and one if i is a solid at T.

In method III, (Figure 2c), the temperature of solid IC is raised from the liquidus temperature to the melting temperature. The solid is then melted, forming the stoichiometric liquid, which is then separated into the pure liquid elements. Finally, the pure liquid components are cooled from the melting temperature of the compound to the liquidus temperature. If the melting temperatures of the pure elements are above the liquidus temperature, this last step again involves subcooling the pure elements. The equation for the standard state chemical potential difference resulting from the sequence in Figure 2c is

$$\theta_{IC} = \frac{T_m}{T} \ln a_I^{sl}(T_m) a_C^{sl}(T_m) + \frac{\Delta S_f^o(I)}{R} \left[\frac{T_m}{T} - 1 \right]$$

$$+ \frac{1}{RT} \int_T^{T_m} \int_T^{T_m} \frac{\Delta C_p^o(IC)}{T} dT^2 \qquad (9)$$

where $a_i^{sl}(T_m)$ is the activity of component i in the stoichiometric liquid at the compound melting temperature, $\Delta S_f^o(IC)$ is the entropy of formation of compound IC from the pure liquids at their melting temperature, and $\Delta C_p^o(IC)$ is the difference in the heat capacities of the pure liquids and the solid compound.

Each of the above three methods employs a different data base. Most of the property values required for the evaluation of θ_{IC} in Equations 7-9 have been experimentally determined for III-V systems and these three relationships can be used as a test for thermodynamic consistency. The first method, Equation 7, is most reliable at or near the binary compound melting temperature. As the temperature is lowered below the melting one, uncertainties in the extrapolated stoichiometric liquid heat capacity and component activity coefficients become important. The second method, Equation 8, is limited to the temperature range in which an experimental determination of ΔG^o is feasible (e.g., high temperature galvanic cell). Method II is also valuable for "pinning down" the low temperature values of θ_{IC}. Method III is the preferred procedure when estimating solution model parameters from liquidus data. Since the activity coefficients of the stochiometric liquid components are evaluated only at the melting temperature, the temperature dependence of θ_{IC} is explicit in this expression.

An Example: Ga-Sb. As an example, the reduced standard state chemical potential was evaluated by each method for the Ga-Sb system. The melting temperature of GaSb is low (985K) relative to other III-V systems and the vapor pressure at this temperature is small when compared to the arsenides and phosphides. Because of the experimental accessibility, values for each of the thermochemical properties required to evaluate θ_{GaSb} with Equations 7-9 have been measured at various temperatures and often by several investigators. The reduced standard state chemical potential for GaSb was calculated by Equations 7 and 9 with the experimental property values selected to give maximum and minimum values for θ_{GaSb}. These upper and lower experimental bounds are shown in Figure 3. The thermochemical property values selected to give these bounds are summarized in Table I.

The dotted line in Figure 3 shows the value of θ_{GaSb} calculated using Equation 8 and the measurements of $G_f^o(GaSb)$ by Abbasov et al. (19). The value of θ_{GaSb} at 298K shown in Figure 3 (circle) was also calculated using Equation 8 with the value of $\Delta G_f^o(GaSb, 298 \text{ K})$ taken from Lichter and Sommelet (20).

A fourth method by which θ_{IC} can be determined from experimental results is an application of Equations 3 and 4. In the binary limit ($X_{AC} = 1$, $x_A = 1 - x_C$) these equations can be solved for θ_{IC} to give

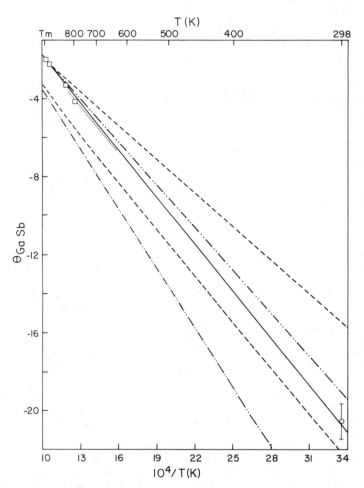

Figure 3.   Values of θ_{GaSb} versus reciprocal temperature. -----, upper and lower bound, Equation 7; ....., Equation 8, ΔG_f^0(GaSb) from ref. (19); 0, Equation 7, ΔG_f^0(GaSb) from ref. (20); —·—, upper and lower bound, Equation 9; ———, recommended values of θ_{GaSb}; □, Equation 10, a_{Ga} from ref. (6), T-x^l from ref. (21), enthalpy of mixing from ref. (22).

Table I. Summary of Ga-Sb Thermochemical Property Values
Used in Equations 7 and 9

Property	Upper Bound	Ref.	Lower Bound	Ref.
$a_{Ga}^{sl}(1003K)a_{Sb}^{sl}(1003K)$	0.134	6	0.0293	14
$T_m(GaSb)$ K	975	7	998	15
$\Delta H_m(GaSb)$ kJ/mole	50.21	8	66.94	16
ΔH_{mix}^{sl} J/mole	-854	9	-1071	17
$\Delta C_p(GaSb)$ J/mole K	5.699	10	2.895	16
$\Delta S_f^o(GaSb,T_m)$ J/mole K	-23.24	11,12	-28.16	10,13
$\Delta C_p^o(GaSb)$ J/mole K	$3.24-9.29 \times 10^{-4}T$	12,13	$33.35-21.68 \log T$	13,18

$$\theta_{IC} = \ln[a_I(T,x_I^1)a_C(T,x_I^1)] = \ln[a_I(T^*,x_I^1)a_C(T^*,x_I^1)]$$

$$- \int_{T^*}^{T} \frac{\Delta\overline{H}_I(x_I^1) + \Delta\overline{H}_C(x_I^1)}{RT^2} dT \tag{10}$$

Here, $a_I(T,x_I^1)a_C(T,x_I^1)$ is the activity product at the liquidus temperature, T, or at some measurement temperature, T^*, and liquidus composition, x_I^1, and $\Delta\overline{H}_i(x_I^1)$ is the relative partial molar enthalpy for component $i = I,C$ at the liquidus composition. The data base required for this final procedure is the phase diagram and the activity product at the liquidus temperature and composition (or an isothermal activity product and the liquid phase enthalpy of mixing). This method requires binary mixture information and will give two values of θ_{IC} at each liquidus temperature, one from each side of the compound, which are identical given a consistent data set. This procedure was applied to four measurements of $a_{Ga}(1003K)$ of Anderson et al. (6) along with the interpolated phase diagram of Maglione and Potier (21) and the enthalpy of mixing results of Gambino and Bros (22). The results are plotted in Figure 3 with square symbols.

As displayed in Figure 3, there exists a large variation in the values which can be assigned to θ_{GaSb}. An examination of the results from all four procedures, though, suggests an appropriate temperature dependence of θ_{GaSb}. The solid line in Figure 3 represents the values of θ_{GaSb} selected for this study.

Determination of Stoichiometric Liquid Component Activities. The calculation of the reduced standard state chemical potential difference by Equation 7 or 9 in the absence of knowledge of the thermodynamic properties of the liquid phase necessitates the imposition of a solution model to represent the product of the component activities in the stoichiometric liquid. In Equation 7 the activity product is required as a function of temperature, while Equation 9 requires that it be calculated only at the melting temperature of the binary compound. When Equation 7 or 9 is used in the calculation of the phase diagram, a solution model is also required to determine the Γ_{IC} term in Equations 3 and 4. Representing θ_{IC} by Equation 7 and determining the activity product with a solution model has been the procedure used almost exclusively in the literature to interpolate, extrapolate and predict III-V phase diagrams (23).

In order to test the effect of using a solution model to calculate values of θ_{IC}, the Non-Random Two-Liquid (NRTL) equation (24) was used in conjunction with Equation 7 or 9 to fit Ga-Sb data sets consisting of the liquidus temperature alone, liquidus temperature and enthalpy of mixing, liquidus temperature and isothermal Ga activity, and all three types of data combined. Using the parameters determined from the data reduction, values of θ_{GaSb} as a function of reciprocal temperature were calculated and compared with the recommended values given in Figure 3. The measured values of liquidus temperature, activity of Ga, and enthalpy of mixing used in the fit are those selected by Aselage et al. (25); these data sets have been shown to be thermodynamically consistent by Anderson et al. (6).

For the first case, Equation 7 was used in which the activity product of the stoichiometric liquid at the liquidus temperature of interest was calculated from the NRTL equation. The enthalpy of fusion and melting temperature for the compound as well as the heat capacity difference, however, were specified. The values chosen for these properties were those recommended by Chang et al. (4). Figure 4 compares the calculated values of θ_{GaSb} for each data set with the recommended values (solid line). In these data reductions the NRTL equation had four adjustable parameters and the non-randomness factor was fixed at -0.001. The results given in Figure 4 are an extreme example of the possible discrepancy between the values of θ_{IC} calculated by this procedure and the recommended values. The ability of the parameter estimates obtained in these fits to reproduce the data sets is within the reported experimental error for each data set. Thus, errors in the extrapolated values of θ_{IC} are canceled by errors in the calculated values of Γ_{IC}. Using values of θ_{IC} determined in this way to extrapolate data sets in temperature or to predict multicom-

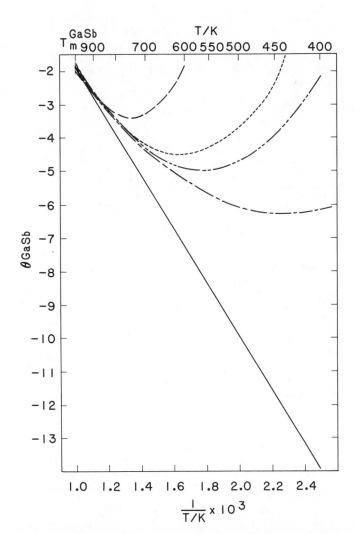

Figure 4. Values of θ_{GaSb} versus reciprocal temperature as calculated by the NRTL equation and Equation 7. Reduction of data set: — — —, liquidus temperature; -- -- --, liquidus temperature and Ga activity; — -- —, liquidus temperature and enthalpy of mixing; -----, liquidus temperature, enthalpy of mixing and Ga activity; ———, recommended values of θ_{GaSb}.

ponent phase diagrams and solution behavior would give unreliable
results.

For the second case, Equation 9 was used with Equations 3 and
4 to reduce the same binary Ga-Sb data sets. The stoichiometric
liquid mixture activity product was fixed at the melting temperature by the NRTL model parameters. The remaining thermodynamic
properties found in Equation 9 were assigned the values recommended by Chang et al. (4).

The results of these calculations are presented in Figure 5.
The temperature dependence of θ_{GaSb} calculated with any of the
data sets is the same as that for the selected values of θ_{GaSb}.
This is a result of the fact that Equation 9 requires the calculation of the activity product only at the compound melting temperature. A reasonably close prediction of the activity product by
the solution model should give a good representation of the temperature dependence. Equations 3 and 4 for a binary system at the
melting temperature reduce to

$$\theta_{IC} = \ln a_I^{sl}(T_m)a_C^{sl}(T_m) \qquad (11)$$

Since the melting temperature is included in the data set for each
case studied, a close prediction of the activity product at the
melting temperature is expected, particularly if component activities are also included in the data set.

Determination of Γ_{IC}

Equilibrium between a pseudobinary III-V solid solution and a ternary liquid solution is described by Equations 3 and 4. By the
methods presented in the previous section, the determination of
the reduced standard state chemical potential change, θ_{IC}, can
proceed in a reliable manner. The other term contained in Equations 3 and 4 is Γ_{IC}, and its determination is discussed here.

Liquid Solution Behavior. The component activity coefficients in
the liquid phase can be addressed separately from those in the
solid solution by direct experimental determination or by analysis
of the binary limits, since $\gamma_{IC} = 1$. Because of the large amount
of experimental effort required to study a ternary composition
field and the high vapor pressures encountered in the arsenide and
phosphide melts, a direct experimental determination of ternary
activity coefficients has been reported only for the Ga-In-Sb system (26). Typically, the available binary liquidus data have been
used to fix the adjustable parameters in a solution model with θ_{IC}
determined by Equation 7. The solution model expression for the
activity coefficient has been used not only to represent the component activities along the liquidus curve, but also the
stoichiometric liquid activities needed in Equation 7. The ternary melt solution behavior is then obtained by extending the
binary models to describe a ternary mixture without additional
adjustable parameters. In general, interactions between atoms in
different groups exhibit negative deviations from ideal behavior

while interactions between atoms in the same group show positive deviations.

A number of attempts have been made to use the simple solution model to represent the solution nonidealities in binary (23,29-32) and ternary (23,33-41) III-V systems. In the simple solution model, the integral Gibbs excess energy is given in terms of an interaction energy $\omega(T)$ by the equation

$$G^{xs} = \omega(T) \; x_A(1-x_A) \tag{12}$$

where $\omega(T)$ is usually given a linear temperature dependence, a+bT.

The binary phase diagrams have been fitted well using this model, although different values of the interaction parameter were necessary for the composition range on either side of the compound for the highly asymmetric Al-Sb system (34). In spite of a wide variety of parameters reported for many of the binary systems, the liquid phase thermodynamic properties derived from any of the parameter sets are in variance with available measurements of these properties. For example, the enthalpy of mixing predicted from a fit of the liquidus is always positive, whereas the available experimental data all show negative values. It is anticipated that the mixing enthalpy is indeed negative in all III-V binary systems, due to the attractive nature of the III-V interaction. Similarly, the values of the interaction coefficients derived from the liquidus fits are in poor agreement with those derived from vapor pressure measurements in the arsenide and phosphide systems (29-31). Knobloch (42) and Peuschel et al. (43) have obtained somewhat better agreement with the use of Krupkowski's formalism for the activity coefficient, while Panish (32) has suggested that agreement between the two sets of parameters may be obtained by adjusting the pure component vapor pressures of phosphorus and arsenic. The discrepancies between the calculated and experimental thermodynamic properties have been discussed by Stringfellow (41), who concluded that while the simple solution model is a useful tool for the calculation of phase diagrams, it is unable to represent the other thermodynamic properties.

A number of other models have been used in conjunction with Equation 7 to calculate the binary phase diagrams. Among these are the quasichemical equation (35,44), a truncated Margules equation (45,46), Darken's formalism (47,48), and various forms of the chemical theory, in which associated liquid species are postulated and some assumptions are made about the physical interactions between the species (49-51). Several of these studies have considered the liquid phase properties as well as the liquidus in the parameter estimation (45,46,51).

Solid Solution Behavior. Attempts to calculate ternary phase diagrams with the use of the available binary simple solution parameters have met with fair success. It is necessary in performing such calculations to assign a value to the solid solution activity coefficient in Equation 6. This has usually been accomplished by a fit of the pseudobinary phase diagram, in which the liquid phase mole fraction of the Group III elements and Group V elements are required to each sum to 0.5, or by setting the solid solution activity coefficient equal to one. The latter approach is only useful in systems such as $Ga_xAl_{1-x}As$ in which the lattice mismatch is nearly zero. The general results of such cal-culations show that the calculated ternary liquidus isotherms are in fair agreement with experimental data, and are insensitive to the choice of a particular set of parameters. The calculated ter-nary solidus isotherms, however, appear to be more sensitive to the values of the interaction parameters, and the agreement is fair to poor for ternary III-V systems. Several investigators (33,35,38) have adjusted the values of the binary parameters to obtain a better representation of the ternary system.

As a specific example, Gratton and Woolley (39) have measured solidus isotherms in the Ga-In-Sb system by annealing samples in the two-phase field followed by rapid quenching. The solid con-centrations were then determined by x-ray analysis. While the parameter set chosen by these authors as well as several others (23,37,38) showed reasonable agreement with the available liquidus isotherms, the agreement with the solidus isotherms was poor in each case. In addition, each of the parameter sets predicts a positive enthalpy of mixing over the entire composition region, whereas the experimental measurements (52,53) indicate that the enthalpy of mixing is negative for most compositions.

Aselage and Anderson (26) have used a binary data base that included experimental values of enthalpy of mixing and component activities in addition to liquidus temperature to estimate binary simple solution parameters. The predicted ternary mixture proper-ties and liquidus temperature were in excellent agreement with experimental measurements. Figure 6 compares the predicted 873K tie lines (dashed lines) to the experimental measurements (solid lines) of Gratton and Woolley (39). The agreement shown between the calculated and experimental solidus is reasonably good and represents a significant improvement over previous attempts to calculate the Ga-In-Sb phase diagram using only binary parameters.

One of the main difficulties with using the pseudobinary phase diagram as a data base for estimating the solid solution properties is that it represents only the high temperature behavior. For most applications, the lower temperature portion of the phase diagram is important (e.g., solidus isotherms in the 700-900K temperature range for analysis of liquid phase epitaxy, the prediction of miscibility gaps in the solid solution). The temperature dependence of the solid solution Gibbs excess energy is found to be sensitive to the solution model and method of data reduction used. As an example, Chang et al. (54) studied the $Ga_xIn_{1-x}Sb$ pseudobinary system. The $Ga_xIn_{1-x}Sb$ liquid mixture was treated either as a ternary mixture of Ga, In and Sb, the thermo-dynamic properties being estimated with binary parameters, or as a binary mixture of GaSb and InSb, the thermodynamic properties cal-culated from the simple solution model with the parameters estimated from a fit of the pseudobinary phase diagram. For both descriptions of the liquid mixture the simple solution equation

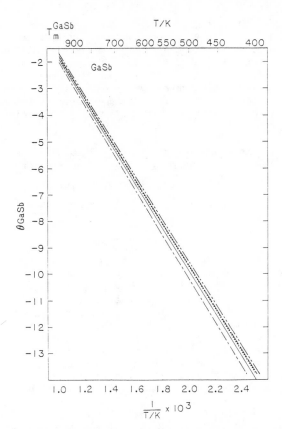

Figure 5. Values of θ_{GaSb} versus reciprocal temperature as calculated by the NRTL equation with Equation 9. Reduction of data set: — — —, liquidus temperature; — · —, liquidus temperature and Ga activity; — – – —, liquidus temperature and enthalpy of mixing; -----, liquidus temperature, Ga activity and enthalpy of mixing; ———, recommended values of θ_{GaSb}.

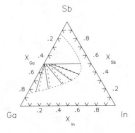

Figure 6. Ga-In-Sb 873K tie lines. -----; simple solution, ———; experimental results (<u>39</u>).

was used to model the solid solution behavior and parameters were estimated from a fit of the pseudobinary phase diagram. Both treatments of the liquid phase gave standard deviations in the liquidus and solidus temperatures within the experimental uncertainty. The variation of the Gibbs excess energy of the solid solution with temperature, however, was in opposite directions for the two different treatments of the liquid phase. It follows that the low temperature segment of the ternary phase diagram is predicted differently for each assumption.

An Example: Γ_{AlSb} in the Al-Ga-Sb System. The importance of the Γ_{IC} term in calculating ternary phase diagrams can be made apparent by examining the distribution coefficient, $K_{IC} \equiv \dfrac{x_{IC}}{x_I}$. Dividing Equation 4 by Equation 3 gives the simple result:

$$K_{IC} = x_C \, \Gamma_{IC} \, \exp(-\theta_{IC}) \qquad (13)$$

Since θ_{IC} is not a function of composition, the slope of a plot of K_{IC} versus x_C along an isotherm is proportional to Γ_{IC}. This plot can be compared to the straight line that is produced when the value of Γ_{IC} is assumed equal to unity, which occurs when either the liquid and solid mixtures are both ideal solutions or the deviation in the liquid phase ($\gamma_I \gamma_C$) is the same as that in the solid (γ_{IC}). Shown in Figure 7 is the AlSb distribution coefficient plotted as a function of the Sb liquid phase mole fraction in the Al-Ga-Sb system at 823K. The experimental values of K_{AlSb} were taken from the solidus and liquidus measurements of Dedegkaev et al. (55) and Chang and Pearson (56). The dashed line represents the distribution coefficient calculated from Equation 13 with the value of $\Gamma_{AlSb} = 1$ and the values θ_{AlSb} selected by Chang et al. (4). The solid line is the result of evaluating Equation 13 with the value of Γ_{AlSb} again unity but with the values of θ_{AlSb1} determined from Equation 7 and the following property values: $a_{Al}^{sl}(T)a_{Sb}^{sl}(T) = 0.25$, i.e. ideal solution, $\Delta H_m(AlSb) = 82.0$ kJ/mole, $T_m(AlSb) = 1338K$ and $\Delta C_P(AlSb) = 10.83$ J/mole K. It is observed from Figure 7 that, first, the value of Γ_{AlSb} is not unity and, second, that there exists a partial cancellation of the composition dependence in the liquid phase activity coefficient product by that found in the solid solution. The second observation suggests that the liquid and solid solution model selection process should be insensitive with respect to liquidus and solidus data alone. Indeed, the assumption of ideal solution behavior in both phases closely predicts the correct distribution coefficient, yet experimental measurements of the solution thermochemical properties clearly indicate moderate negative deviations from ideal behavior.

An "experimental" value for Γ_{IC} can be assigned by solving Equation 13 for Γ_{IC} and requires values for θ_{IC}, based on the experimental properties discussed previously, and liquidus and solidus data. Shown in Figure 8 is a plot of $1-\Gamma_{AlSb}$ versus x_{Sb} along several isotherms. It is seen that Γ_{AlSb} is nearly

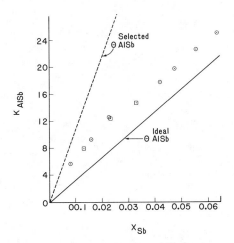

Figure 7. The AlSb distribution coefficient versus the Sb liquid phase mole fraction at 823K. -----, Γ_{AlSb} = 1; ———, AlSb = 1 and $a_{Al}^{sl}(T)a_{Sb}^{sl}(T)$ = 0.25; O, measured values (<u>56</u>); □, measured values (<u>55</u>).

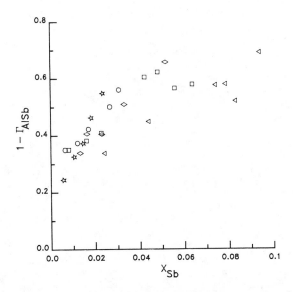

Figure 8. Values of $1-\Gamma_{AlSb}$ calculated with Equation 13 versus the liquid phase mole fraction at various temperatures. The calculation used the recommended value of θ_{AlSb} (<u>4</u>) and the phase diagram determinations: ☆, 778K (<u>55</u>); ◇, 825K (<u>55</u>); ◁, 873K (<u>55</u>); O, 773K (<u>56</u>); □, 823K (<u>56</u>).

independent of temperature in the temperature range studied. In addition, Γ_{AlSb} varies with composition only at low values of x_{Sb}, at which the experimental uncertainty in Γ_{AlSb} is large. The partial cancellation in both the composition and temperature dependence of Γ_{AlSb} is largely responsible for the ability of rather simple solution models, which incorrectly represent liquid solution properties, to represent ternary phase diagrams.

Summary and Conclusions

A consistent formalism by which ternary III-V phase diagrams could be calculated was developed. It was shown that the ternary calculation could be separated into a determination of a standard state term, θ_{IC}, which is a function only of temperature, and a term, Γ_{IC} which represents the ratio of the nonidealities in the liquid phase to those in the solid solution.

Four different methods were presented to determine the reduced standard state chemical potential change and applied to the Ga-Sb system. It is common practice to use Equation 7 and a solution model representing the stoichiometric liquid activities to determine θ_{IC}. The solution model parameters are then estimated from a fit of the binary phase diagram. It has been shown that this procedure can lead to large errors in the value of θ_{IC}. The use of Equation 9, however, gave the correct temperature dependence of θ_{IC} and the inclusion of activity measurements in the data base replicated the recommended values of θ_{IC}.

The determination of Γ_{IC} was examined by first considering the liquid solution behavior and then the solid mixture properties. The liquid phase properties are typically determined by using a solution model to interpolate between the binary limits. In general, the use of only the binary phase diagrams in the data base for model parameter estimation does not give good values for the ternary liquid mixture properties. The solid solution behavior is normally determined from an analysis of the pseudo-binary phase diagram. Extrapolation of the solid solution behavior determined in this manner to lower values of temperature should be undertaken with caution.

Important to calculation of the ternary phase diagram, however, is the ratio of the liquid phase component activity coefficients to the solid component one. As illustrated for the Al-Ga-Sb system, the temperature and composition variation of the liquid phase component activity coefficients is largely cancelled by that of the solid component activity coefficient.

Acknowledgments

The authors wish to express their appreciation to Dr. John O'Connell for many helpful discussions. Support of this work by the Ceramics Division of the National Science Foundation under Grant DMR-8012684 is gratefully acknowledged.

Literature Cited

1. Hurle, D.T.J. J. Phys. Chem. Solids 1979, 40, 613.
2. Edelin, G.; Mathiot, D. Phil. Mag. B 1980, 42, 95.
3. Brebrick, R.F. Met. Trans. 1971, 2, 1658.
4. Chang, K.M.; Aselage, T.L.; Anderson, T.J. To be submitted
    J. Electrochem. Soc.
5. Vieland, L.J. Acta Met. 1963, 11, 137.
6. Anderson, T.J.; Aselage, T.L.; Donaghey, L.F. J. Chem. Ther-
    modynamics 1983, 15, 927.
7. Welker, H. Physica 1954, 20, 893.
8. Schottky, W.F.; Bever, M.B. Acta Met. 1958, 6, 320.
9. Yazawa, A.; Kawashima, T.; Itagaki, K. J. Jap. Inst. Met.
    1968, 32, 1288.
10. Hultgren, R.; Desai, P.D.; Hawkins, D.T.; Gleizer, M.; Kel-
    ley, K.K. In "Selected Values of the Thermodynamic Properties
    of Binary Alloys"; Am. Soc. Metals: Metals Park, Ohio, 1973.
11. Sirota, N.N.; Yushkevich, N.N. Khim. Svyas v Poluprov i
    Tverd. Telakh Inst. Fiz. Tverd. Tela i Polyprov., Akad. Nauk
    Belorusok. SSR 1965, 122.
12. Lundin, C.E.; Pool, M.J.,; Sullivan, R.W.   Denver Research
    Inst. Final Report No. AFORL-63-156, 1963.
13. Hultgren, R.; Desai, P.D.; Hawkins, D.T.; Gleizer, M.; Kel-
    ley, K.K.; Wagman, D.D. In "Selected Values of the Thermo-
    chemical Properties of the Elements"; Am. Soc. Metals: Metals
    Park, Ohio, 1973.
14. Gerasimenko, L.N.; Kirichenko, I.V.; Lozhkin, L.N.; Mora-
    chevskii, A.G. Metal i Oksidyne Pokrytiya, Korrosiya Metal i
    Issled v Obl. Electrokhim. Akad.   Nauk SSSR Otd. Shch, i
    Tekhn. Khim. Sb. Statei 1965, 236.
15. Cunnell, F.A.; Edmond, J.T.; Richards, J.L. Proc. Phys. Soc.
    (London) 1954, B67, 848.
16. Cox, R.H.; Pool, M.J. J. Chem. Eng. Data 1967, 12, 247.
17. Predel, B.; Stein, D.W. J. Less Common Met. 1971, 24, 391.
18. Maslov, P.G.; Maslov, Yu.P. Chemical Bonds in Semiconductors
    and Solids 1972, 3, 191.
19. Abbasov, A.S.; Nikol'skaya, A.V.; Gerasimov, Ya.I. Dolk.
    Akad. Nauk SSSR 1964, 156, 1399.
20. Lichter, L.D.; Sommelet, P. Trans. Met. Soc. AIME 1969, 245,
    99.
21. Maglione, M.H.; Potier, A. J. Chim. Phys. 1968, 65, 1595.
22. Gambino, M.; Bros, J.P. J. Chem. Thermodynamics 1975, 7,
    443.
23. Panish, M.B.; Ilegems, M. In "Prog. in Solid State Chem.";
    Reiss, H.; McCaldin, J.O., Eds.; Pergamon Press: New York,
    1972; Vol. 7, p. 39.
24. Renon, H.; Prausnitz, J.M. A.I.Ch.E. J. 1968, 14, 135.
25. Aselage, T.L.; Chang, K.M.; Anderson, T.J. Accepted CALPHAD
    1985.
26. Aselage, T.L.; Anderson, T.J. Accepted High Temp. Sci.
    1985.
27. Hall, R.N. J. Electrochem. Soc. 1963, 110, 385.

28. Sol, N.; Clariou, J.-P.; Linh, N.T.; Moulin, M.   J.   Crystal Growth 1974, 27, 225.
29. Thurmond, C.D.   J. Phys. Chem. Solids 1965, 26, 785.
30. Arthur, J.R.   J. Phys. Chem. Solids 1967, 28, 2257.
31. Ilegems, M.; Panish, M.B.; Arthur, J.R.   J. Chem. Thermo- dynamics 1973, 5, 5046.
32. Panish, M.B.   J. Crystal Growth 1974, 27, 6.
33. Wu, T.Y.; Pearson, G.L.   J. Phys. Chem. Solids 1972, 33, 409.
34. Ilegems, M.; Pearson, G.L.   Proc. 1968 Symp. on GaAs The Institute of Physics and the Physical Society, London, 1969.
35. Joullie, A.; Gautier, P.; Monteil, E.   J. Crystal Growth 1979, 47, 100.
36. Osamura, K.; Inoue, J.; Murakami, Y.   J. Electrochem. Soc. 1972, 119, 103.
37. Blom, G.M.; Plaskett, T.S.   J. Electrochem. Soc. 1971, 118, 1831.
38. Joullie, A.; Dedies, R.; Chevrier, J.; Bougnot, G.   Rev. Phys. Appl. 1974, 9, 455.
39. Gratton, M.F.; Woolley, J.C.   J. Electrochem. Soc. 1978, 125, 657.
40. Tomlinson, J.L.   Naval ocean Systems Center, Tech. Note 386, San Diego, CA, 1978.
41. Stringfellow, G.B.   J. Crystal Growth 1974, 27, 21.
42. Knobloch, G.   Krist. und Tech.   1975, 10, 605.
43. Peuschel, G.-P.; Knobloch, G.; Butter, E; Apelt, R.   Crys. Res. Tech. 1981, 16, 13.
44. Stringfellow, G.B.   J. Electrochem. Soc.   1970, 117 1301.
45. Kaufman, L.; Nell, J.; Taylor, K.; Hayes, F.   CALPHAD 1981, 5, 185.
46. Liao, P.-K.; Su, C.-H.; Brebrick, R.F.; Kaufman, L.   CALPHAD 1983, 7, 207.
47. Antypas, G.A.   J. Electrochem. Soc.   1970, 117, 1393.
48. Antypas, G.A.   J. Electrochem. Soc.   1970, 117, 700.
49. Osamura, K.; Predel, B.   Trans. Japan Inst. Metals 1977, 18, 765.
50. Szapiro, S.   J. Phys. Chem. Solids 1980, 41, 279.
51. Liao, P.-K.; Su, C.-H.; Tung, T.; Brebrick, R.F.   CALPHAD 1982, 6, 141.
52. Chang, K.M.; Coughanowr, C.A.; Anderson, T.J.   Submitted to Chem. Eng. Comm.   1984.
53. Ansara, I.; Gambino, M.; Bros, J.P.   J. Crystal Growth 1976, 32, 101.
54. Vecher, A.A.; Voronova, E.I.; Mechkovskii, L.A.; Skoropanov, A.S.   Russ. J. Phys. Chem.   1974, 48, 584.
55. Dedegkaev, T.T.; Kryukov, I.I.; Lideikis, T.P.; Tsarenkov, B.V.; Yakovlev, Yu.P.   Sov. Phys. Tech. Phys.   1978, 23, 350.
56. Cheng, K.Y.; Pearson, G.L.   J. Electrochem. Soc.   1979, 126, 1992.

RECEIVED March 12, 1985

Preparation of Device-Quality Strained-Layer Superlattices

R. M. Biefeld

Sandia National Laboratories, Albuquerque, NM 87185

A strained-layer superlattice (SLS) consists of very thin (~ 500 Å) alternating layers of lattice mismatched semiconductors. When the layers are thin enough the mismatch between the materials is totally accommodated by strains within the layers and no misfit defects are generated. Thus device quality material can be prepared from a variety of mismatched semiconductors by using SLSs. Also, it has been shown that the SLS band gap can be varied independently of the lattice constant over a continuous range of lattice constants by the proper choice of composition and layer thickness. Utilizing the unique structural, optical, and electrical properties of SLSs, high-quality heterostructures can be prepared for use in a variety of device applications. The quality of SLSs has been characterized by current-voltage, electroluminescence and mobility measurements. These materials have been shown to be of suitable quality for use in heterojunction devices. A variety of devices such as photodiodes, field effect tranistors and lasers have been prepared from SLSs. The characterization of some of these devices is discussed.

In the past, heterojunction devices have been largely limited to lattice matched systems (1). This is primarily because of the generation of misfit dislocations during the growth of mismatched materials. These dislocations severely degrade the heterojunction device performance. Recently it has been found that if sufficiently thin layers of mismatched materials are grown epitaxially, the formation of misfit dislocations can be minimized or entirely tirely eliminated (2,3). The layer thickness, at which dislocations begin to form depends on the lattice mismatch and the properties of the materials in the superlattice (4). For thin layers the mismatch is entirely taken up by straining the layers so that the lattice constants match up in the planes parallel to the growth surface. A series of alternating layers of these very thin, lattice-mismatched materials, when taken as a whole, are designated strained-layer superlattices (SLSs).

0097–6156/85/0290–0297$06.00/0
© 1985 American Chemical Society

A schematic of a strained-layer superlattice which illustrates the tetragonal distortion that cubic layers undergo is presented in Figure 1. If alternating, mismatched layers are grown, the strain accommodates the mismatch in such a way that the layers are successively in tension and compression which results in equal lattice parameters for both layers parallel to the interfaces, a . The Poisson effect causes the layers in compression to expand in the direction perpendicular to the growth surface and the layers in tension to contract in the same perpendicular direction. This gives rise to a tetragonal distortion for cubic materials and two different lattice parameters in this perpendicular direction, $a_\perp$. The strain in the layers has been found to act as a barrier to the propagation of misfit dislocations (3). Thus, not only are the layers themselves free of dislocations, but if bulk materials which are latticed matched to the SLS are grown on top of the SLS they can also be prepared free of misfit dislocations.

It has been shown that these strained-layers have unique optical and electronic properties (5,6). The band gap in these SLSs has been found to depend on the strain in the layers as well as the composition and layer thickness. Also, the energy of the band gap and the lattice constant of the SLS can be varied independently (7). Therefore, the use of SLSs either as a structural barrier to the propagation of misfit dislocations in mismatched systems or as "bulk" materials with unique properties, greatly increases the range of materials' properties which can be used in heterostructure devices.

This paper will describe some of the recent structural, optical and electronic characterizations of SLSs which indicate that these materials are of device quality. The preparation and properties of selected devices will also be presented.

Preparation and Structural Characterization

Strained-layer superlattices were first prepared by A. E. Blakeslee in 1971 using a vapor phase reactor (8). The structures consisted of alternating layers of GaAs and $GaAs_{0.5}P_{0.5}$ grown on (100)GaAs. Examination of these SLSs by x-ray diffraction, electron microscopy and etch pit studies showed that SLSs could be prepared with low dislocation densities and that they would act as barriers for the propagation of threading dislocations (3,4,9). More recently, SLSs have been prepared by metal organic chemical vapor deposition ($GaP/GaAs_{1-x}P_x$) and molecular beam epitaxy ($GaAs/Ga_{1-x}In_xAs$) (6,10-12). Both of these growth techniques have been shown to be capable of producing very thin and compositionally uniform epitaxial layers with very sharp interfaces (11,13,14). The quality of the MOCVD interfaces is shown in the transmission electron micrograph in Figure 2 and by the lattice image in Figure 3. Figures 2 and 3 indicate that very few defects are present at the interfaces and that the crystal growth of alternate layers proceeds without the generation of new misfit dislocations.

X-ray diffraction patterns of an SLS allow for the determination of the layer thickness, composition and strain in the layers (14-16). A typical x-ray diffraction pattern for a superlattice grown directly on a GaP substrate is shown in Figure 4. The layer thickness is related to the observed d-spacings by the following,

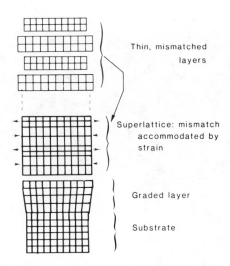

Figure 1.   Schematic of a strained-layer superlattice (SLS) showing the tetragonal distortion from the original cubic layers.

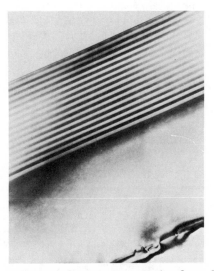

Figure 2.   Transmission electron micrograph of an SLS with alternating layers of GaP and $GaAs_{0.18}P_{0.82}$ each 225A. Misfit dislocations are evident only in the lower right hand corner at the interface between the substrate and buffer.

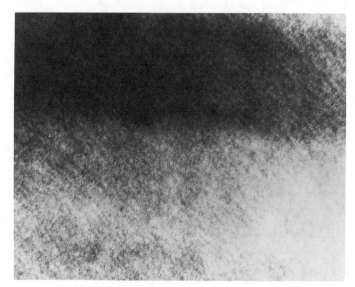

Figure 3.   High resolution transmission electron microscope (111)
            lattice image of the same structure in Figure 2.   The
            light and dark areas are the GaP and $GaAs_{0.18}P_{0.82}$
            layers which are horizontal.   The light lines which are
            diagonal are the (111) lattice planes.   The distance
            between the planes is $\sim 3.16\text{\AA}$.

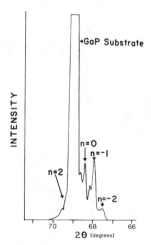

Figure 4.   X-ray diffraction pattern of an SLS grown directly on a
            GaP substrate.   The presence of two peaks for each
            satellite is due to the α_1 and α_2 components of the
            CuK_2 radiation.

$$N = nd/[4(d_0-d)] \tag{1}$$

for the (400) reflection where N is the number of unit cells with a lattice constant of $4d_0$ in the superlattice unit cell, n is the order of a particular satellite reflection, d is the interplanar spacing determined from the nth reflection and d_0 is the average lattice spacing (n=0). The sharpness of the diffraction lines is a qualitative indication of the compositional sharpness and strain uniformity in the layers. A quantitative fit of a calculated curve to the experimental data allows one to estimate the uniformity of the layers. When this is done for MOCVD layers in the $GaP/GaAs_{1-x}P_x$ system, variations in the composition and the layer uniformity are estimated to be less than five percent of the measured values (16). Double-crystal x-ray diffraction studies allow for the determination of depth profiles of parallel and perpendicular strain. Results of these studies indicate that the lattice misfit in SLSs is elastically accommodated by tetragonal distortions in the alternating layers (16).

Ion channeling experiments have been carried out on both MOCVD grown $GaP/GaAs_{1-x}P_x$ and MBE grown $GaAs/Ga_{1-x}In_xAs$ SLSs (17-19). Results of these studies support both the presence of tetragonal distortions in the SLS layers and the strain accommodation of misfit in SLSs. The ion scattering results indicate that the crystalline quality of the SLS is as good as that of the substrate.

Optical and Electronic Properties

Although the first SLSs were prepared in the 1970's (8), the optical and electrical characterization of different SLS structures was carried out only recently (6,7,10-14,21-23). Both photoluminescence and photocurrent spectroscopy have been used to characterize the band structure of SLSs in the $GaP/GaAs_{1-x}P_x$ and $GaAs/In_{1-x}Ga_xAs$ systems (21-26). Spectra measurements (21-23) were found to be in good agreement with the band structure predicted by tight-binding and effective-mass (Kronig-Penny) models (5,23). The efficient photoluminescence that was observed is further evidence that the SLSs are of high crystalline quality.

An interesting feature of the luminescent properties of the SLSs is the independent variability of the bandgap and lattice constant (7,23). This result is illustrated in Figure 5 which shows the variation of the bandgap for a series of SLSs in the $GaAs_xP_{1-x}$ system. It should be noted that the band gap can be varied between the values shown for the thick well limit and those of the bulk $GaAs_xP_{1-x}$ alloy by the appropriate choice of layer compositions. A total of four SLS properties can be varied independently for ternary SLSs with two layers per period (27). This unique variability should allow one to tailor-make SLSs for specific device applications.

The band structure of the SLS is determined not only by the composition of the layers and their thicknesses but also by the strain in the layers and quantum size effects (5). The strain affects the energy of the band minima, and splits certain degenerate levels in both the conduction and valence band. The splitting which results from the strain can also alter the effective mass of the holes in the SLSs (20).

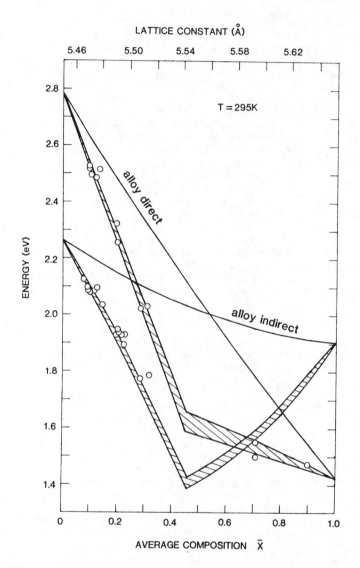

Figure 5. Variation of SLS band gap with layer composition and
           thickness at 300K.   Lines are for theoretical calcula-
           tions and circles are from photocurrent and photolumin-
           escence data.   The shaded area is for equal layer
           thicknesses from 60Å, top line, to the thick well
           limit, bottom line.

Another of the consequences to the band structure of a superlattice in the (100) GaP/GaAs$_{1-x}$P$_x$ system is to fold some of the
indirect conduction minima onto the direct (Γ) minimum (5,20).
This zone folding effectively creates a direct gap SLS material
from bulk materials which have indirect band gaps. As depicted in
Figure 6, the absorption coefficients for these materials are
enhanced over that for indirect materials. However, this absorption is less than that for direct gap materials due to the spatial
separation of the electrons and holes within the SLS (5,20,28).
This spatial separation is a result of the band offsets in the
GaP/GaAs$_{1-x}$P$_x$ system which causes the conduction band minimum to
lie in the GaP layer and the valence band maximum in the GaAs$_{1-x}$P$_x$
layer.

Optical techniques have also been used to measure the minority
carrier diffusion lengths in both n- and p-type GaP/GaAs$_{1-x}$P$_x$ SLSs
(29). The measured diffusion lengths perpendicular to the SLS
interfaces were ~ 0.1 µm and the parallel diffusion lengths
were ~ 1.5 µm. The parallel diffusion lengths are similar to
those measured in bulk GaP. The small perpendicular diffusion
length is consistent with the existence of potential barriers in
both the conduction and valence bands which inhibit transport
normal to the SLS interfaces.

Hall effect measurements have been carried out on a wide
variety of SLS samples at different temperatures (12,15,30). Some
of these results are presented in Table I. The mobilities in
these materials are as good as those for similar bulk alloys. In
the case of the GaAs/In$_{0.2}$Ga$_{0.8}$As modulation doped structure, a
significant enhancement (μ (35K) ~ 35,000 cm^2/V-sec) over the
bulk mobility has been observed (30). This enhancement is similar
to those which have been observed in lattice matched (GaAs/Al$_{1-x}$Ga$_x$As)
superlattices and is a sensitive test for the presence of ionized
impurities and other charged defects (20,31). Shubnikov-de Haas
oscillations have been found in similar GaAs/In$_{1-x}$Ga$_x$As SLSs (32).
These studies are strong evidence that the misfit dislocation
densities in SLSs are small enough so that high quality optoelectronic devices could be made from SLS material.

Fabricated Devices

In order to fabricate devices in the SLSs, one must be able to
successfully dope the SLSs both n- and p-type. This has been
achieved and diodes have been prepared in the GaP/GaAs$_{1-x}$P$_x$ and
GaAs/In$_{1-x}$Ga$_x$As systems (33,34). The I-V response of a typical
diode for the GaP/GaAs$_{1-x}$P$_x$ system is shown in Figure 7. The
diode in Figure 7 consists of an n-type (100) GaP substrate, an
n-type (5 x 10^{17} cm^{-3}), 1.2 µm GaAs$_{.1}$P$_{.9}$ buffer layer, an n-type
(9 x 10^{16} cm^{-3}), 0.43 µm, 34 layer GaP/GaAs$_{0.2}$P$_{0.8}$ SLS and a
p-type (6 x 10^{17} cm^{-3}), 0.18 µm, 14 layer GaP/GaAs$_{0.2}$P$_{0.8}$ SLS.
The diode has a reverse leakage current of -279 pA at -5.0 V, an
n-factor of ~ 2 and a turn on voltage of 1.8 V. Thus, in spite of
the existence of a large number of interfaces within the active
region of this diode, it exhibits reasonable diode characteristics.
Similar high quality diodes have been reported for GaAs/In$_{1-x}$Ga$_x$As
(34).

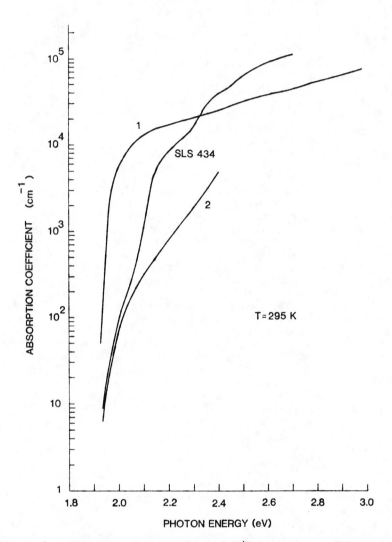

Figure 6.   Absorption spectra for a GaP/GaAs$_{0.44}$P$_{0.56}$ SLS.   Curve
1 is direct-gap GaAs and curve 2 is indirect-gap GaP
shifted to energy gaps of 1.92 eV equal to that of the
SLS.

Table I.   Electrical Data from Hall Measurements
on Selected SLS Samples

Sample	T K	Carrier Concentration cm^{-3}	Mobility cm^2/V-sec
GaP/GaAs$_{.08}$P$_{.92}$	300	1.4×10^{17} (P)	55
GaP/GaAs$_{.14}$P$_{.86}$	300	4.9×10^{17} (n)	183
GaAs/GaAs$_{.78}$P$_{.22}$	300	1.8×10^{17} (P)	209
	77	1.0×10^{17} (P)	1,240
GaAs/GaAs$_{.66}$P$_{.34}$	300	1.2×10^{19} (n)	1,930
	77	5.2×10^{18} (n)	2,660
GaAs/In$_{0.2}$Ga$_{0.8}$As*	300	1.1×10^{17} (n)	7,000
	77	1.1×10^{17} (n)	31,000
	4	1.1×10^{17} (n)	32,000
GaAs/In$_{0.2}$Ga$_{0.8}$As	300	3.5×10^{18} (P)	100

* Modulation doped

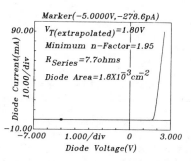

Figure 7.   Current-voltage trace of a GaP/GaAs$_{0.2}$P$_{0.8}$ grown junction diode.

These diodes have also been operated as photodetectors
(33,34). The wavelength response of the $GaP/GaAs_{1-x}P_x$ diode described above is shown in Figure 8. The quantium efficiency of
this diode is greater than 40% at ~ 400 nm. This efficiency was
achieved without optimizing the structure and with no antireflection coating. Similar results are found for $GaAs/In_{1-x}Ga_xAs$
diodes. SLS junctions in both these material systems also emit
electroluminescence under forward bias which indicates that minority carrier injection is taking place (20). These results again
indicate that device quality SLSs can be prepared.

A field-effect transistor has been prepared using a modulation
doped $GaAs/In_{0.2}Ga_{0.8}As$ SLS as the channel of the device (35). The
device consisted of a p-type GaAs substrate, a p-type (4×10^{18}
cm^{-3}), 1 μm $In_{0.1}Ga_{0.9}As$ bufffer, an n-type (1×10^{17} cm^{-3}),
0.51 μm, 34 layer $GaAs/In_{0.2}Ga_{0.8}As$ SLS with a 4 layer n^+-SLS
contacting top layer. Gate control of the transistor was accomplished by an Aℓ Schottky diode on the top and a p-n junction
diode on the bottom. Figure 9 gives the room temperature dc
common-source output characteristics for an FET with a 2.5 μm
gate length and the upper and lower gates shorted externally.
The intrinsic transconductance (g_{mi}) for this device was calculated
to be 120 mS/m ($V_{DS} = 4V$) at room temeprature. At 77K the device
characteristics improved by ~ 60% and g_{mi} increased to 190 mS/mm
(35).

A variety of photopumped laser structures have been fabricated
by other researchers in $GaAs/In_{1-x}Ga_xAs$, $GaAs/GaAs_{1-x}P_x$ and
$GaAs_{1-x}P_x/In_{1-x}Ga_xAs$ SLSs which were grown by MOCVD (24,36). Continuous, photopumped lasing action has been observed in these
materials at 300K. Although these lasers were found to degrade
after various periods of time, the demonstration of stimulated
emission is strong evidence for the high quality of SLS materials.
The mechanism for degradation in these SLS lasers is not completely
understood (20).

An ion-implantation technology is being developed for SLSs
(37). High quality p-n junction diodes have been fabricated by Be^+
implantation in $GaP/GaAs_{1-x}P_x$ SLSs. Activation was achieved by
annealing the implanted wafers at 825°C for 10 minutes in an over-
pressure of AsH_3 and PH_3 in H_2. X-ray diffraction indicated that
the SLS survived this procedure. The diodes formed by this procedure had n-factors of ~ 2.0 and a reverse leakage current of 1.5
$\times 10^{-7}$ A/cm^2 at -10 V. The severe conditions which the SLSs are
exposed to during implantation and annealing to achieve activation
is evidence of their thermal and structural stability.

The above devices, junction diodes, FET, lasers and ion
implanted diodes, demonstrate that high quality devices can be
made from SLS materials. Use of the unique properties of SLSs
should greatly expand the areas of applications for heterostructure
devices.

Acknowledgments

The author would like to acknowledge the many helpful discussions with I. J. Fritz, P. L. Gourley, D. R. Myers, G. C. Osbourn
and T. E. Zipperian which made the writing of this paper possible.
This work performed at Sandia National Laboratories supported
by the U.S. Department of Energy under Contract Number
DE-ACO4-76-DP00789.

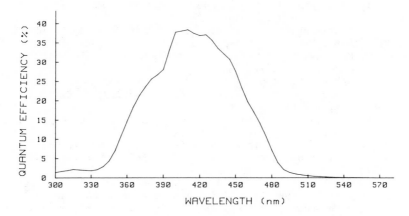

Figure 8.   Photoresponse of the diode shown in Figure 7.

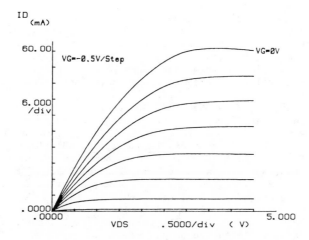

Figure 9.   Room temperature output characteristics of a modulation
            doped GaAs/In$_{0.2}$Ga$_{0.8}$As SLS FET which shows drain
            current versus drain source voltage for -0.5V steps in
            the gate source voltage.

Literature Cited

1. A. G. Milnes and D. L. Feucht, "Heterojunctions and Metal-Semiconductor Junctions," Academic Press, New York, NY (1972).
2. J. H. van der Merwe, J. Appl. Phys., 34, 117 (1963).
3. J. W. Matthews and A. E. Blakeslee, J. Crystal Growth, 32, 265 (1976).
4. J. W. Matthews, A. E. Blakeslee, and S. Mader, Thin Solid Films, 33, 253 (1976).
5. G. C. Osbourn, J. Appl. Phys., 53, 1586 (1982).
6. G. C. Osbourn, R. M. Biefeld and P. L. Gourley, Appl. Phys. Lett., 41, 172 (1982).
7. R. M. Biefeld, P. L. Gourley, I. J. Fritz and G. C. Osbourn, Appl. Phys. Lett., 43, 759 (1983).
8. A. E. Blakeslee, J. Electrochem. Soc., 118, 1459 (1971).
9. A. Segmueller and A. E. Blakeslee, J. Appl. Cryst., 6 19 (1973).
10. P. L. Gourley, R. M. Biefeld, G. C. Osbourn, and I. J. Fritz, Proceedings of 1982 International Symposium on GaAs and Related Compounds, Inst. of Physics, Conf. Ser. 65, Bristol, England (1982), p. 248.
11. I. J. Fritz, L. R. Dawson, G. C. Osbourn, P. L. Gourley, and R. M. Biefeld, Proceedings of 1982 International Symposium on GaAs and Related Compounds, Inst. of Physics Conf. Ser. 65, Bristol, England (1982), p. 241.
12. I. J. Fritz, L. R. Dawson and T. E. Zipperian, J. Vac. Sci. Technol., B1, 387 (1983).
13. R. M. Biefeld, Ind. Eng. Chem. Prod. Res. Dev., 21, 525 (1982).
14. R. M. Biefeld, G. C. Osbourn, P. L. Gourley, and I. J. Fritz, J. Electronic Mater., 12, 903 (1983).
15. R. M. Biefeld, G. C. Osbourn, P. L. Gourley, I. J. Fritz, and T. E. Zipperian, Proceedings III-V Opto-electronics Epitaxy and Device Related Processes Symposium, Electrochem. Soc., Pennington, NJ (1983), p. 217.
16. V. S. Sperisou, M. A. Nicolet, S. T. Picraux and R. M. Biefeld, submitted to Appl. Phys. Lett. (1984).
17. S. T. Picraux, R. M. Biefeld, L. R. Dawson, G. C. Osbourn, and W. K. Chu, J. Vac. Sci. Technol., B1, 687 (1983).
18. S. T. Picraux, L. R. Dawson, G. C. Osbourn, R. M. Biefeld, and W. K. Chu, Appl. Phys. Lett., 43, 1020 (1983).
19. W. K. Chu, J. A. Ellison, S. T. Picraux, R. M. Biefeld, and G. C. Osbourn, Phys. Rev. Lett., 52, 125 (1984).
20. G. C. Osbourn, "Thin Films and Interfaces Symposium," Eds. J. E. E. Baglin, E. R. Campbell, and W. K. Chu, Materials Research Soc. (1984), to be published.
21. I. J. Fritz, R. M. Biefeld and G. C. Osbourn, Solid State Commun., 45, 323 (1983).
22. P. L. Gourley and R. M. Biefeld, J. Vac. Sci. Technol., B1, 383 (1983).
23. G. C. Osbourn, Phys. Rev. B, 27, 5126 (1983).
24. M. J. Ludowise, W. T. Dietze, C. R. Lewis, N. Holonyak, K. Hess, M. D. Camras, and M. A. Nixon, Appl. Phys. Lett., 42, 257 (1983).
25. W. D. Laidig, J. W. Lee, P. K. Chang, L. W. Simpson, and S. M. Bedair, Electronic Materials Conf., University of Vermont,

June, 1983: The Metallurgical Society of AIME, Warrendale, PA (1983), Abstract E-3.

26. J. Y. Marzin and E. V. K. Rao, Appl. Phys. Lett., $\underline{43}$, 560 (1983).

27. G. C. Osbourn, J. Vac. Sci. Technol., $\underline{B1}$, 379 (1983).

28. G. C. Osbourn, J. Vac. Sci. Technol., $\underline{21}$, 469 (1982).

29. P. L. Gourley, R. M. Biefeld, T. E. Zipperian, and J. J. Wiczer, to be published, Appl. Phys. Lett. (1984).

30. I. J. Fritz, L. R. Dawson, and T. E. Zipperian, Appl. Phys. Lett., $\underline{43}$, 846 (1983).

31. R. Dingle, H. L. Stormer, A. C. Gossard, and W. Wiegmann, Appl. Phys. Lett., $\underline{33}$, 365 (1978).

32. J. E. Schirber, I. J. Fritz, L. R. Dawson, and G. C. Osbourn, Phys. Rev. B, $\underline{28}$, 2229 (1983).

33. R. M. Biefeld, T. E. Zipperian, P. L. Gourley, and G. C. Osbourn, Electronic Materials Conf., University of Vermont, June, 1983:
Metallurgical Society of AIME, Warrendale, PA (1983), Abstract E-2.

34. L. R. Dawson, G. C. Osbourn, T. E. Zipperian, J. J. Wiczer, C. E. Barnes, I. J. Fritz, and R. M. Biefeld, Molecular Beam Epitaxy Workshop, Georgia Inst. Technol., October, 1983, Abstract TA7.

35. T. E. Zipperian, L. R. Dawson, G. C. Osbourn, and I. J. Fritz, International Electron Devices Meeting, December, 1983: IEEE, Piscataway, NJ (1983), p. 696.

36. M. P. Camras, J. M. Brown, N. Holonyak, Jr., M. A. Nixon, and R. W. Kaliski, J. Appl. Phys., $\underline{54}$, 6183 (1983).

37. D. R. Myers, T. E. Zipperian, R. M. Biefeld, and J. J. Wiczer, International Electron Devices Meeting, December, 1983: IEEE, Piscataway, NJ (1983), p. 700.

RECEIVED January 23, 1985

Wafer Design and Characterization for Integrated-Circuit Processes

R. Schindler, D. Huber, and J. Reffle

Wacker Chemitronic, Gesellschaft für Elektronik-Grundstoffe GmbH, Postfach 1140, D-8263 Burghausen, Federal Republic of Germany

The new generations of electronic circuits with scaled geometries and very large scale integration put stringent requirements to the starting silicon material. State of the art crystal growth techniques allow all relevant process parameters to be precisely controlled. As a consequence the performance of wafers in IC processing becomes much more predictable and reproducable, and the starting material can be tailored to a specific device manufactoring process for improved device performance and yields.

I) Design and Device Physics Considerations

Some of the requirements for building functioning electronic devices are simply dictated by the device specifications and by physical laws. A high voltage diode, for instance, for 500 V reverse bias voltage, is a device where a big volume is electrically active. To allow for high reverse bias voltages the resistivity of the material has to be high, which in turn means that the depletion layer is large, too. For the diffusion length of minority carriers to be large the material has to be very pure and perfect crystalline silicon. Thus the device specification for high reverse bias voltage and small leakage current determines substrate resistivity, thickness of the starting wafer, and, as will be clear later, growth process of the silicon single crystal.

In contrast to power devices, which essentially are volume devices, most of the integrated circuits are small signal devices. Here only little volume in depth is electrically active. With shrinking dimensions and supply voltages dropping devices become more and more two dimensional with their electrically active depth being in the order of just a few microns. This feature allows materials characteristics in the surface-near electrically active area to be different from the bulk of the wafer.

It is known that dynamic random access memories are built on substrates with different resistivities according to some designer's choice. Given the resistivity, the depth of the depletion layer depends on the voltage applied only (fig.1), while the diffusion length of carriers is determined by the density of recombination centers N_T (trap density). The leakage current I_R of a diode (fig.2) is given by

$$I_R = I_{Gen} + I_{Diff.}$$

with I_{Gen} = generation current and $I_{Diff.}$ = diffusion current.

The generation current is (after /1/)

$$I_{Gen} = \frac{1}{2} q \frac{n_i}{\tau} W A_j \approx \frac{1}{2} q n_i \sigma v_{th} N_T \, sech\left\{\frac{E_i - E_t}{kT}\right\} A_j$$

n_i = intrinsic carrier concentration
τ = effective lifetime for minority carriers
W = depth of depletion layer
σ = capture cross section
v_{th} = thermal velocity
N_T = trap density
A_j = junction area
E_t = energy level of trap
E_i = intrinsic Fermi level

It is that part of the leakage current, which is due to generation of carriers within the space-charge region. If the trap centers are located near the intrinsic Fermi level E_i, then τ_{eff} will be practically independent of temperature. The generation current then has the same temperature dependence as n_i. The generation-current component is dependent on the magnitude of the applied reverse bias - at higher biases W is larger, more centers are included within the depletion region, and the generation current increases in proportion to W.

The diffusion current in reverse bias is given by /1/

$$I_{Dif} = q \frac{n_i^2 D}{L_D N_A} A_j$$

D = diffusion coefficient for carrier
N_A = acceptor concentration (doping level)
L_D = diffusion length

It is that part of the leakage current which is due to diffusion of minority carriers to the space charge region.

The contributions of the generation current and the diffusion current, respectively, can be seperated by means of their temperature dependence. As $I_{Gen} \sim n_i \sim exp\left\{- E_g/2kT\right\}$ (E_g = Energy of bandgap) and $I_{Dif} \sim n_i^2 \sim exp\left\{ - E_g/ kT\right\}$, the diffusion current distribution is predominant at elevated temperatures ($\gtrsim 60°C$). In

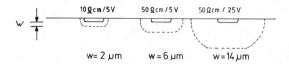

Figure 1. Depth of depletion layer for different substrate resistivity and supply voltages.

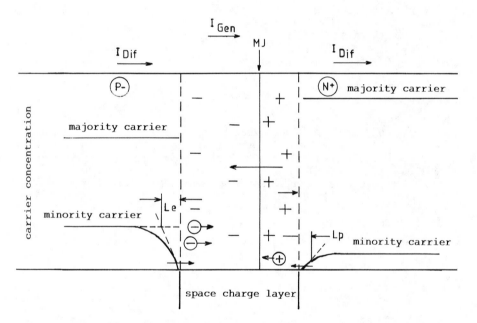

MJ   metalurgical junction

L_e, L_p diffusion length of minority carriers

Figure 2. Sketch of currents contributing to leakage in p-n
junctions (after /35/).

fig.3 the MOS storage time is plotted as a function of temperature. The MOS storage time very often is deduced from C-t-measurements, where the capacitance of a capacitor is measured as a function of time. The discharge of the capacitor is due to a leakage current. It can be seen easily that at lower temperature the generation current is dominating, at higher temperature the diffusion current.

It is absolutely important to control both the trap density inside the space charge layer as well as the trap density in the vicinity of the junction. Thus the specification of the leakage current in a device design will determine the minimal diffusion length of the material necessary at the end of the device fabrication process.

II) Defects

As a consequence of the physics of electronic devices the control of the trap density N_T (and the diffusion length L_D) is the key for successful device manufacturing.

a) Grown-in crystal imperfections

Crystal growth today provides material with essentially no grown-in crystal lattice defects like dislocations or stacking faults. Also, the chemical purity of the wafers with the exception of oxygen and carbon is so high that the diffusion length is in the order of 400 µm, i.e. N_T is very small. The overall impurity concentration is in the order of $\lesssim 10^{11}$ at/cc. In CZ grown silicon, however, oxygen is incorporated during crystal growth and remains in supersaturation for relevant process temperatures. In the course of device processing, oxygen related defects will grow, which may bear some influence on the leakage current.

b) Process induced defects

Oxygen precipitates themselves do not necessarily degrade device performance and yield. The detrimental effects are rather caused by crystal defects generated during oxygen precipitation. Which defects are formed depends on numerous parameters like the oxygen content, oxidation temperature, ambient etc.

For instance, for high oxygen contents and processing temperature of $\approx 1000°C$, punched out dislocation loops are formed. They are emitted from a growing oxygen precipitate, while only the early stages of bulk stacking fault growth are observed (fig.4) /3,4/.

At 1100°C, however, predominantly stacking faults are formed close to the surface and in the bulk /5,6/ (fig.5,6). Their length depends on oxidation time and atmosphere. Si self-interstitials formed in supersaturation at the oxidation front are absorbed by the stacking faults close to the wafer surface and do not migrate into the bulk of the wafer, while the stacking faults deep in the bulk grow due to incorporation of the interstitials emitted from

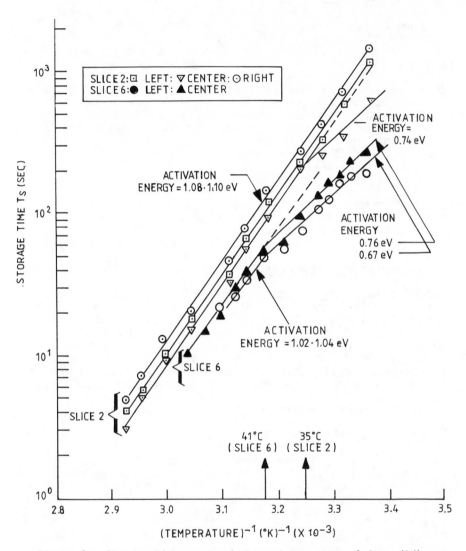

Figure 3.   Storage time versus inverse temperature (after /2/).

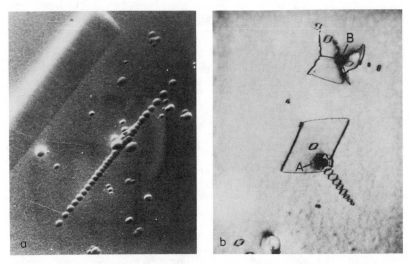

Figure 4. Punched-out dislocations from oxygen precipitates.
a) Etch pits after 3 min Secco etch on cleavage face.
b) TEM micrograph (from /4/).

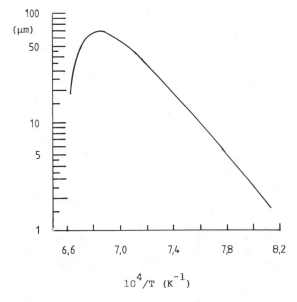

Figure 5. Stacking fault length vs. inverse temperature from /3/.

growing oxygen precipitates. Additionally, bulk stacking faults and punched-out dislocations may react and lead to total annealing of the punched-out dislocations. The stacking fault formation both in the bulk and near the surface is governed by heterogeneous nucleation. The supersaturation of self-interstitials typically is not high enough to allow for homogeneous nucleation of oxidation-induced stacking faults, since the critical nuclei sizes for self-interstitial agglomerates are fairly high /7/. Several different nucleation agents have been identified. The most important ones are surface damage, residual damage from ion-implantation, and wafer contamination with transition-metals (e.g. Ni,Fe, Cr).

These rather mobile metals may be introduced during processing. Epitaxy, ion-implantation, and high-temperature annealing (e.g. drive-in process) are particularly notorious sources for metal contamination /8/. At low temperatures these metals become supersaturated and tend to precipitate. Due to their high diffusivity, they are able to precipitate at energetically favorable sites like the wafer surface or the strain field of other defects. Surface near precipitates provide nuclei for (OSF). Their tendency to decorate other defects is the basis of intrinsic gettering. Metallic impurities also decorate the OSF they nucleated and thereby impart harmful "electrical activity" to otherwise fairly docile stacking faults (fig.7).

Oxidations in chlorine containing ambients are frequently used methods to reduce the influence of metallic impurities during high temperature processing.

Gettering of unwanted impurities is a very important part of device processing. The most efficient getter cycle is the so-called "phosphorous getter" /9/. Phosphorous diffusion is, according to one model, accompanied by generation of phosphorous-vacancy complexes, which are possible sinks for the metallic atoms. The gettering of metallic impurities is driven by the gain in the free energy obtained by the dissolution of the impurity in the phosphorous-vacancy complexes. For a vacancy-trapped impurity, the electrostatic interaction energy with the phosphorous atom is so high that this complex is stable during further processing. In the highly doped phosphorous region the metallic impurity is electrically harmless, as Auger recombination already limits minority carrier lifetime.

Extrinsic and intrinisic gettering mechanisms rely on the interaction of impurities with dislocations which are intentionally introduced into the backside of the wafer (extrinsic) or precipitation of oxygen in the bulk of the wafer (intrinsic). This "Cottrell" gettering is due to the large strainfields associated with dislocations which attract the impurities. It has been found /10/ that the efficiency of the Cottrell gettering is related to the core structure, degree of dissociation, etc. of the dislocations. This explains why freshly formed dislocations seem to be more efficient than "aged" dislocations, since dislocations tend to relax into low energy configuration by extensive local atom re-ordering during processing.

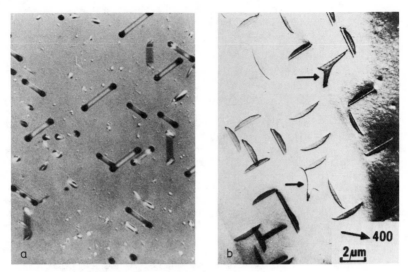

Figure 6. Stacking faults
a) Cleaveage face after 3 min Secco etch, (110).
b) TEM micrograph, (100) (From /4,11/).

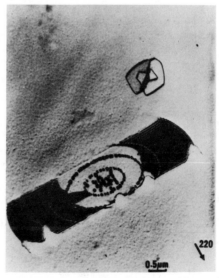

Figure 7. Decorated Stacking faults (from /4,11/).

In Si wafer processing, there are primarily two causes for stress apart from oxygen precipitation. The first one are temperature gradients. The second causes are surface-near layers with structural properties different from that of the Si-substrate, e.g. oxides, nitrides, poly-Si, or doped layers. If these stresses in a wafer grow beyond a critical, temperature-dependent level, plastic deformation will occur by movement of dislocations. Dislocations will lead mostly to catastrophic device failure /11,12/.

The critical stress needed for homogenous nucleation of dislocations is considerable larger than that for dislocation propagation or multiplication. In order to tolerate fairly high stress levels during processing, the heterogenous nucleation of dislocations has to be avoided /13/ (fig.8).

Maximum stress, for instance, is produced at film edges; a particularly critical system is the nitride-edge during local oxidation /14,15/ (fig.9). Not only does the nitride-layer inherently create stress in the Si-substrate (which only is partially relieved by the buffer oxide), but the oxide-growth introduces even more stress because the nitride-cover at the edge has to bend upwards to accomodate the volume expansion of the growing oxide (bird's beak phenomenon) /16/.

Whereas film-edge stresses are very localized and tend to create high densities of dislocations in rather small volumes, other sources of stress may afflict the whole volume of the wafer, such as stresses introduced by temperature gradients.Then defects in any location within the wafer may act as dislocation sources, in particular SiO_x-precipitates and stacking faults. This may result in warpage or slippage /13/ which renders the whole wafer useless.

c) Oxygen

In CZ grown material oxygen is incorporated into the crystal via the silicon melt in the crucible. The oxygen solubility in silicon /17/ is temperature dependent and at relevant processing temperatures the oxygen is supersaturated in the crystal.

In general terms the precipitation of oxygen can be described as /17/

$$\frac{C(t) - C^*(T)}{C_o - C^*(T)} = \exp\left\{-\frac{t}{\tau}\right\}$$

The left term represents the oxygen concentration normalized to the solubility limit $C^*(T)$ for a given temperature, C_0 denotes the starting concentration of the oxygen in solution. The right hand side gives the precipitation rate with the time constant τ.

For every oxygen atom that is removed into a precipitate, the crystal gains a certain amount of free enthalphy, but at the same time has to spend some energy for the increase in the Si-SiO_x interface area. The net energy balance therefore depends on the precipitate size and temperature. The addition of an oxygen atom to a precipitate smaller than a critical radius r_c will not lead to an energy gain, but to an energy increase; this is illustrated in fig. 10a. The critical radius r_c corresponds to the maximum in energy in fig.10a ; smaller precipitates (called "embryos") tend to shrink. This, however, does not mean that embryos do not exist because at finite temperatures their "energy-levels" will always be populated with a certain probability that is given by the Boltzmann-factor; this is illustrated in fig. 10b. At any given temperature therefore embryos of critical size exist, some of which could grow to large precipitates if diffusion can take place. This is the essence of the homogeneous nucleation theory as it has been applied to 0-precipiation in Si with some success (/19/,/20/).

Very slowly precipitating material may be due to homogeneous nucleation. For this case τ is expected to be in the order of 100 hrs. Heterogeneous nucleation will be similar to homogeneous nucleation, except that the energy levels in fig. 10 are lower. This leads either to more embryos of critical size, to smaller critical sizes, or to both.

There is strong evidence that heterogeneous nucleation does occur during SiO_x-precipitation /21/ at high levels of carbon (2×10^{17} at/cc) or in the presence of small agglomerates of self-interstitials or vacancies. Small values of τ (20 hrs) may reflect heterogeneous nucleation.

It should be noted that in most experiments reported, only thin wafers are used. This implies that the influence of the nearby surfaces is always present. It is therefore possible to interpret experiments in terms of heterogeneous nucleation, although really homogeneous precipitation took place in the ingot.

Since τ reflects in some way the number of nuclei, we expect τ to change during processing, because new nuclei are formed during ramping and push/pull of the wafers. Fig.11 demonstrates this effect clearly. The precipitation rate after repeated oxidations is accelerated as compared to an equivalent single step heat treatment.

Even more complicated is the situation in real processing where different temperature sequences influence τ dramatically. A process simulation with a temperature sequence of 1100°C, 1230°C, 1100°C, 1200°C, and 1100°C yielded precipitation constants of 300 hrs, 5 hrs, 10 hrs, 3 hrs, and 5 hrs, respectively.

The precipitation rate is not only dependent on the number of nuclei, but also on the absolute amount of oxygen within the silicon single crystal. Fig. 12 shows the change of oxygen to be primarily a function of the initial oxygen concentration. This also holds true for material grown with relatively small τ

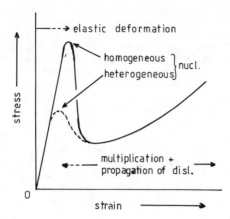

Figure 8.   Idealized stress-strain relation for Si (after /13,15/).

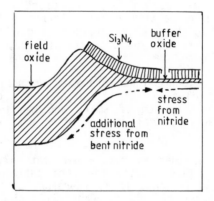

Figure 9.   Sketch of oxide-edge with stress (after /15/).

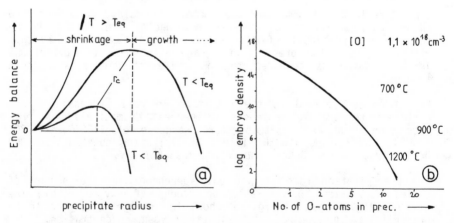

Figure 10. a) Energy balance of the crystal for the formation of a precipitate (after /7/; arbitrary units):
b) Density of embryos as a function of embryo size (after /15,18/).

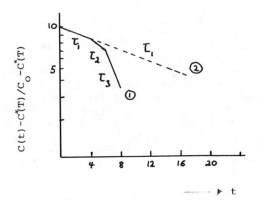

Figure 11. Precipitation rate differences due to push/pull. (1) 3 oxidations, (2) straight oxidation for 16 hrs.

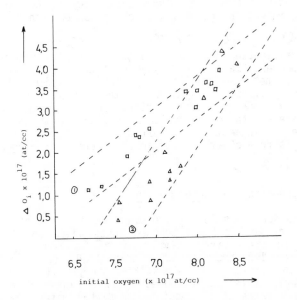

Figure 12. Change of oxygen vs. initial oxygen content in material grown with different growth processes (1) and (2).

In addition to controlling the precipitation rate by means of crystal growth techniques (pullrates, furnace design etc.), the precipitation can be accelerated by annealing of whole ingots or individual wafers. In this annealing step additional nuclei are formed. Theoretically this should take place at temperatures as low as possible to ensure small values for the critical radius r_c. However, it requires also diffusion of oxygen. In the semiconductor industry temperatures in the range of 650°C to 800°C are reported /14/. In this temperature region a certain fraction of embryos larger than r_c will grow. Their absolute number is mostly determined by the embryo-population that was frozen-in during crystal growth. This explains why reproducible precipitation requires starting material which naturally would have homogeneous nucleation only. Fig. 13 gives an example of possibilities for trimming the precipitation behavior.

At the same time outdiffusion of oxygen to the surface plays an important role. The simplest model follows from the solution of the one-dimensional diffusion equation, subjected to the boundary conditions of uniformly initial oxygen concentration C_0 that is subsequently pinned to the oxygen solubility C_s (T) at the surface of the wafer. The resulting oxygen concentration C (x,t) is:

$$C(x,t) = C_s \, erf(u) - C_0 \, erf(u) \; ; \quad u = x/2Dt$$

Several investigations have been published on this experiment /22, 23, 24/. However, they are all based on thermal donor formation of oxygen and do not give the direct profile of oxygen. Also, one of the remaining questions is that for the correct surface concentrations C_s of oxygen in the outdiffusion experiments. The precipitation rate and the outdiffusion of oxygen are the dominant factors in the development of the denuded zone, a zone near the surface which is free of crystallographic defects and oxygen precipitates.

III) N_T Engineering and Wafer Design

All the considerations for influencing the precipitation behaviour so far aim at well defined denuded zones which ensure the trap density N_T in the device active areas to be extremely small and the diffusion length to be well controlled. It is possible to calculate the profile of interstitial oxygen (fig.14) near the surface of the wafers. In connection with material of suitable precipitation rates it is possible to tailor wafers very well to specific device needs. Moreover, it can be engineered in a way that the final device needs can be satisfied for different manufacturing processes /25/. Fig.15 gives an example, how the depth of the denuded zone can be varied due to different preanneals.

Andrews /24/ and more recently Watanabe et al /26/ developed models for the denuded zone formation within the limitations of measurement tools for oxygen profiles. For MOS processing, which essentially has process temperatures around and below 1000°C, a three step process is required to set up a denuded zone. One for the out-diffusion of oxygen, one for nucleation of SiO_x preci-

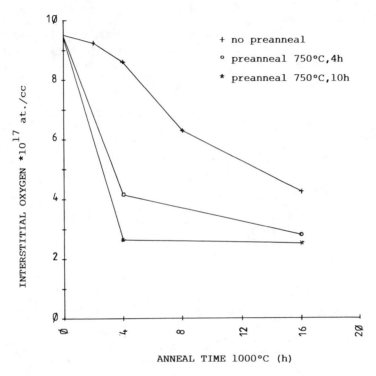

Figure 13. Interstitial oxygen concentration for different preanneal times.

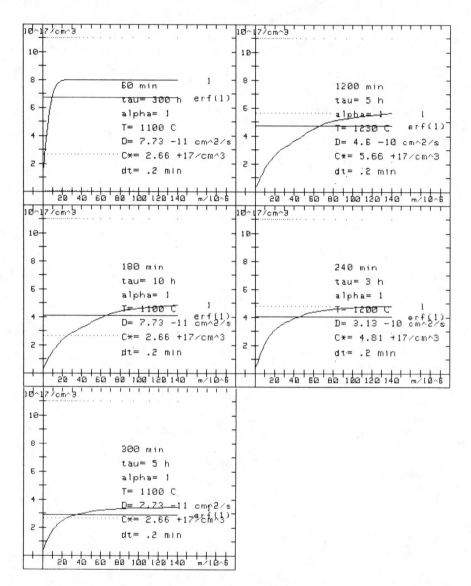

Figure 14. Simulation of interstitial oxygen profile during device processing. The dashed line represents the solubility limit of oxygen at process temperature.

pitation , and one for SiO_x precipitate growth. For bipolar processes with higher temperatures the out-diffusion will take place during initial oxidation and collector diffusion.

With controlled precipitation, intrinsic gettering takes place
during processing. Because of the ability of metallic impurities
to decorate other defects, specifically oxygen precipitates, these
impurities are removed out of the denuded zone into the bulk of
the wafer with beneficial effects for device performance.

Oxygen precipitates are recombination centers for minority carriers.
They then are limiting the diffusion length L_D. A silicon wafer with
a denuded zone and a high concentration of precipitates is electrically an inhomogeneous system.

So L_D is not the same throughout the wafer thickness and may be
characterized by an effective diffusionlength L_D^{eff}. L_D^{eff} may be
calculated by properly taking into account the high lifetime
region close to the surface and the low lifetime region of the
bulk (fig.16).
Applying these considerations to generation lifetime we would
expect a drop in lifetime outside of the DZ (fig.17). This in fact
is observed. Fig.18 shows lifetime-measurements on a wafer, which
after precipitation was slightly wedge shaped polished in order
to vary the depth of the DZ. Subsequently MOS capacitors had been
built and the lifetime measured. The increase in lifetime is due
to the DZ.

Tayloring a denuded zone depth results in tayloring diffusion
length L_D^{eff} and trap density N_T. This has influence on different
aspects for device performance.

Actual device yield in bipolar processing is reported to be greatly improved; if the substrate material is precipitating slowly,
i.e. if the precipitation rate is very well controlled /29,30/.
It is also known that the refresh behavior in MOS DRAMS is improved, if carefully designed DZ's exist at the end of the device
manufacturing process /31/.

As the denuded zone concept relies essentially on a two layer system other applications seem to be possible, which heavily depend
on material with layered differences in lifetime (diffusion length).
One of the areas is in the field of α-particle sensitivity. α
-particles emitted from the device package or from the environment
create electron-hole pairs, which may lead to soft errors in memories. This can be suppressed if the carriers generated recombine at
the oxygen precipitates rather than being able to diffuse to the
memory cells.

Accordingly, the 2-layer-lifetime system can be used to improve
latch-up susceptibility of CMOS devices. Improved isolation techniques like trench isolation prevents lateral diffusion of carriers,
while the vertical diffusion of carriers leads to their fast recombination at precipitates.

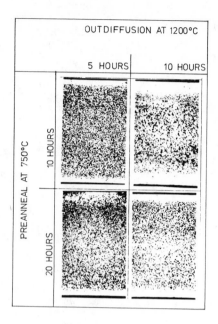

Figure 15. Denuded zone depths for different process conditions.

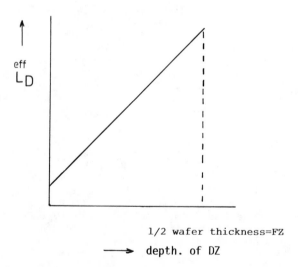

Figure 16. Effective diffusion length vs. depth of denuded zone. Note: There is a geometrical limit of the maximum diffusion length.

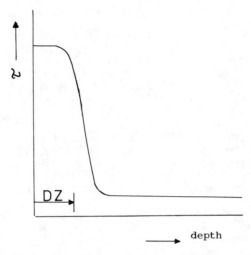

Figure 17. Expected relation between lifetime and depth.

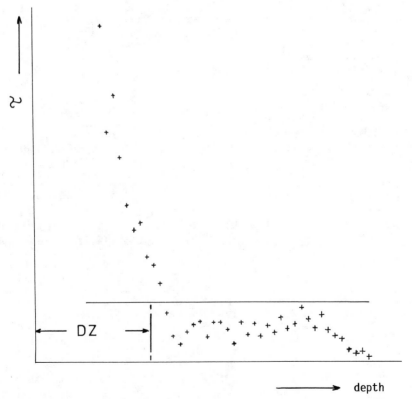

Figure 18. Generation lifetime of capacitors with varying distances to the precipitated area of the wafer.

IV) Wafer characterization

A number of characterization methods for wafers are used today. One group of experiments aims at the purely electronic parameters like DLTS and Photoluminiscence. While DLTS and related measurements give information on the trap energies, methods like surface photovoltage (SPV) or open circuit voltage decay (OCVD) give information on effective diffusion length.

a) SPV
Monochromatic light creates electron-hole pairs in the wafer, which partly diffuses to the surface. They are seperated at the surface space charge layer, which is due to the band bending associated with the surface. This carrier seperation can be measured as surface photovoltage, which is proportional to the amount of carriers separated. Varying the wavelength of the light gives different depths for the electron-hole pair generation. From the diffusion of the carriers to the space charge layer from the depth the diffusion length can be calculated (fig.19).

Fig.20 gives a comparison between cross sectional measurements and SPV measurements, which are non-destructive. In essence, there exists correlation between the measurements of the depth of the denuded zone on the cleavage face and the diffusion length. SPV thus could be developed into a tool for non-destructively measuring DZ depths.

b) MOS generation lifetime
The generation lifetime can be determined by measuring the relaxation of an MOS capacitor to its equilibrium inversion state after being driven from accumulation into deep depletion by a voltage pulse. The lifetime is calculated after Zerbst /32/ using the slope of the straight-line segment of the plot $-d\ (C_0/C^2)/dt$ vs $(C_{EQ}/C-1)$, where C_{EQ} is the equilibrium inversion capacitance, C_0 is the oxide capacitance, and C is the instantaneous device capacitance. The measurement is very time-consuming, since the measurement time is essentially the holding time of the capacitor. A different method was developed by Pierret and Small /33/. There, an initially deep-depleted MOS capacitor is maintained at a constant capacitance by adding charge to the gate to match that building-up in the inversion layer. A constant device capacitance is equivalent to a constant depletion depth and hence a constant active bulk generation region. This kind of analysis is much faster than analysis following the Zerbst proposal. Both methods, however, have in common, that at least two process steps are required: A gate oxide has to be grown and a gate has to be formed.

Fig.21 shows an example for the generation lifetime distribution across a wafer. Note the high resolution of the plot. It reflects the homogeneity of the critical electrical parameters across wafers representing the state of the art of growing and shaping techniques.

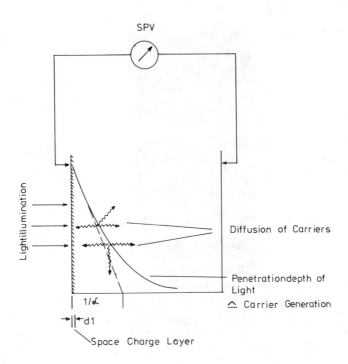

Figure 19. Basic principles of surface photo-voltage measurement.

ELECTRICAL MEASUREMENT

MEASUREMENT OF DIFFUSIONLENGTH (L_D) BY THE
SURFACE PHOTO VOLTAGE (SPV) METHOD

EXAMPLES

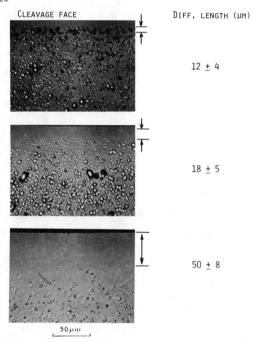

CLEAVAGE FACE   DIFF. LENGTH (µM)

12 ± 4

18 ± 5

50 ± 8

50 µm

Figure 20. Determination of DZ-depth by diffusion-length measurements. The markers to the right of the micrographs represent
the measured diffusion lengths.

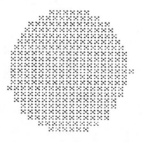

Figure 21. MOS generation lifetime map. Each point represents
a MOS capacitor. The average lifetime is 500 µsec, the standard
deviation is 70 µsec.

Conclusions

In the preceding chapters it has been shown how device performance criteria are influencing the choice of starting material for IC fabrication. Denuded Zones in the material imply the diffusion length in the wafer to be inhomogenous. For integrated circuits this feature allows to improve the device performance and to actually design it into the starting material and the device manufacturing process.

Acknowledgments

The authors are indepted to Drs.B.O. Kolbesen and H.Föll for the TEM micrographs and stimulating discussions, and to Ms. Knörr for assistance in preparing the manuscript.

Literature Cited

/1/ A. S. Grove, "Physics and Technology of Semiconductor Devices", John Wiley, New York, 1967

/2/ J. E. Lawrence and H. R. Huff, "VLSI Electronics-Microstructure Science", N.G. Einspruch ed., Vol.5, Academic Press. N.Y., 1982

/3/ W. Zulehner and D. Huber in "Crystals", H.C. Freyhardt ed., Vol.8, Springer, N.Y., 1982

/4/ B.O. Kolbesen and K.R. Mayer, Final Report, "Crystal defects in large scale IC of silicon", BMFT-NT 665, 1979, Government of the Federal Republic of Germany

/5/ C. Claeys, G. Declerck, R. Van Overstraeten, H. Bender, J. Van Landuyt, and S. Amelinckx, "Semiconductor Silicon 1981", H.R. Huff, R. J. Kriegler, and Y. Takeishi eds., Vol. 81-5, Electrochemical Society, Pennington N.Y., 730, 1981

/6/ C.L. Claeys, "Proceedings of the Third Brazilian Workshop on Microelectronics", Campinas, Sao Paolo, 1981

/7/ H. Föll, U. Gösele, and B.O. Kolbesen, J. Crystal Growth 40, 90 1977

/8/ P.F. Schmidt and C.W. Pierce, J. Electrochemical Society, 128, 630, 1981

/9/ L. Baldi, G.F. Cerotolini, and G. Ferla, J. Electrochemical Soc., 127, 165, 1980

/10/ A. T. Wink, C. J. Werkhoven, and C. V. Opdorp, "Semiconductor Characterization Techniques LP. A. Barnes and G. A. Rozgonyi eds. 259, Electrochemical Society, Pennington, N.Y., 1978

/11/ H. Föll and B.O. Kolbesen, "Semiconductor Silicon 1977", H.R. Huff, B.O. Kolbesen, and T. Abe eds, 740, Electrochemical Society, Pennington, N.Y., 1977

/12/ B.O. Kolbesen and H. Strunk to be published in "VLSI Electronics: Microstrcture Science", N.G. Einspruch ed., Vol.7, Academic Press, N.Y., 1984

/13/ K. Sumino in ref. /5/, 208, 1981

/14/ S.M. Hu, S.P. Kleppner, R.O. Schwenker, and D.K. Seto, J. Appl. Phys. 47, 4098, 1976

/15/ H. Föll, A. Papp., and B.O. Kolbesen, "Technical Proceedings Semiconductor Processing and Equipment Symposium, Semicon Europa 84", Semiconductor and Materials Institute, Mountain View, CA, 67, 1984

/16/ D. Chin, S.Y. Ohm, S. M. Hu, R.W. Dutton, and J.L. Moll, IEEE Electro., Dev. 30, 744, 1983

/17/ R.A. Craven in ref. /5/, 254, 1981

/18/ F.S. Ham, J.Phys.Chem.Solids, 6 335 (1958)

/19/ N. Inoue, K. Wada, and J. Osaka in ref. /5/, 282, 1981

/20/ H.F. Schaake, S.C. Baber, and R.F. Pinizzotto in ref. /5/, 273, 1981

/21/ S. Kishino, Y. Matsushita, M. Kamamori, and T. Iizuka, Jap. J. Appl. Phys. 21, 1982

/22/ S.M. Hu, Appl. Phys. Lett., 36, 561, 1980

/23/ P. Gaworzewski and G. Ritter, Phys. Stat. Sol. (a), 67, 511, 1981

/24/ J. Andrews, Extended Abstracts, 83-1, Electrochem. Society, Pennington, N.Y. 415, 1983

/25/ J. Reffle and D. Huber, Solid State Technology, August 137, 1983

/26/ S. Isomae, S. Aoki, and K. Watanabe, J. Appl. Phys., 55, 817, 1984

/27/ E.T. Nelson, C.N. Anagnostopoulos, J.P. Lavine, and B.C. Burkey, IEEE Electr. Dev. ED 30, 1392, 1983

/28/ C.N. Anagnostopoulos, E.T. Nelson, J.P. Lavine, K.Y. Wong, and D.N. Nichols, IEEE Electron. Dev. ED, 147, 1983

/29/ B. Goldsmith, L. Jastrzebski, and R. Soydan in ref. 22, 486, 1983

/30/ R. Soydan, L. Jastrzebski, and B. Goldsmith, in ref. /22/, 490, 1983

/31/ H. Otsuka, K. Watanabe, H. Nishimura, H. Iwai, and H. Nihira IEEE Electron Devices Letters EDL-3 182, 1982

/32/ M. Zerbst, Zeitschrift für angewandte Physik, 22, 30, 1966

/33/ R.F. Pierret and D.W. Small, ED-20, 457, 1973, IEEE Trans. El. Dev.

/34/ R.F. Pierret, ED 22, 1051, 1975

/35/ T.M. Frederiksen, "Intuitive IC Electronics", McGraw-Hill, N.Y. 1982.

RECEIVED August 1, 1985

Author Index

Subject Index

A

Production by Meg Marshall
Indexing by Susan F. Robinson
Jacket design by Pamela Lewis

Elements typeset by Hot Type Ltd., Washington, D.C.
Printed and bound by Maple Press Co., York, Pa.

RECENT ACS BOOKS

"Organic Phototransformations in Nonhomogeneous Media"
Edited by Marye Anne Fox
ACS SYMPOSIUM SERIES 278; 308 pp.; ISBN 0-8412-0913-8

"Xenobiotic Metabolism: Nutritional Effects"
Edited by John W. Finley and Daniel E. Schwass
ACS SYMPOSIUM SERIES 277; 382 pp.; ISBN 0-8412-0912-X

"Bioregulators for Pest Control"
Edited by Paul A. Hedin
ACS SYMPOSIUM SERIES 276; 540 pp.; ISBN 0-8412-0910-3

"Nutritional Bioavailability of Calcium"
Edited by Constance Kies
ACS SYMPOSIUM SERIES 275; 200 pp.; ISBN 0-8412-0907-3

"Chemical Process Hazard Review"
Edited by John M. Hoffmann and Daniel C. Maser
ACS SYMPOSIUM SERIES 274; 124 pp.; ISBN 0-8412-0902-2

"Dermal Exposure Related to Pesticide Use:
Discussion of Risk Assessment"
Edited by Richard C. Honeycutt, Gunter Zweig,
and Nancy N. Ragsdale
ACS SYMPOSIUM SERIES 273; 530 pp.; ISBN 0-8412-0898-0

"Macro- and Microemulsions: Theory and Applications"
Edited by Dinesh O. Shah
ACS SYMPOSIUM SERIES 272; 502 pp.; ISBN 0-8412-0896-4

"Purification of Fermentation Products:
Applications to Large-Scale Processes"
Edited by Derek LeRoith, Joseph Shiloach, and Timothy J. Leahy
ACS SYMPOSIUM SERIES 271; 200 pp.; ISBN 0-8412-0890-5

"Reaction Injection Molding: Polymer Chemistry and Engineering"
Edited by Jiri E. Kresta
ACS SYMPOSIUM SERIES 270; 302 pp.; ISBN 0-8412-0888-3

"Materials Science of Synthetic Membranes"
Edited by Douglas R. Lloyd
ACS SYMPOSIUM SERIES 269; 496 pp.; ISBN 0-8412-0887-5

"The Chemistry of Allelopathy: Biochemical Interactions Among Plants"
Edited by A. C. Thompson
ACS SYMPOSIUM SERIES 268; 466 pp.; ISBN 0-8412-0886-7

"Rubber-Modified Thermoset Resins"
Edited by Keith Riew and J. K. Gillham
ADVANCES IN CHEMISTRY SERIES 208; 370 pp.; ISBN 0-8412-0828-X

"The Chemistry of Solid Wood"
Edited by Roger M. Rowell
ADVANCES IN CHEMISTRY SERIES 207; 588 pp.; ISBN 0-8412-0796-8